ENGINEERING & DESIGN APPLICATIONS

TECHNOLOGY
Engineering & Design

James LaPorte, Ph.D.

Mark Sanders, Ph.D.

McGraw Hill Glencoe

New York, New York   Columbus, Ohio   Chicago, Illinois   Woodland Hills, California

Glencoe

The **McGraw·Hill** Companies

TABLE OF CONTENTS

INTRODUCING ENGINEERING & DESIGN APPLICATIONS

This workbook presents nine exciting projects that are designed to guide you as you apply the concepts of engineering and design through hands-on activities. Originally developed with a grant from the National Science Foundation (NSF), these expanded and updated projects provide in-depth applications of science, technology, engineering, and math.

Each real-world project includes sections that introduce you to the project: **Design Brief, Challenge, Criteria, Constraints,** and **Enginering Design Process.** Background information, sub-activities, and review questions are grouped in sections titled **Research & Design, Modeling,** and **Evaluation.** Together, the projects focus on the six types of technology:

- Transportation (Projects 6 and 7)
- Communication (Project 9)
- Energy and Power (Project 4)
- Construction (Projects 2, 3, and 5)
- Bio-Related (Project 8)
- Manufacturing (Projects 1 and 5)

Pick-and-Place Robot

Design Brief

Robots are used for a variety of production processes, such as auto assembly (see **Fig. 1-1**) and electronics manufacturing. In addition, chemical production plants use robots to help produce hazardous materials. Since the chemicals they make can cause injuries to employees and others, safe handling methods are important. For this reason, robots handle hazardous materials.

Challenge

Design and construct a model of a robotic device that will transport hazardous materials in chemical drums from the ground to a loading dock in a minimum amount of time.

Fig. 1-1 Robots have many production applications, inlcuding assembly and handling hazardous materials.

Criteria

- You will be given a 35 mm film container to represent a chemical drum. The drum will be filled with water to represent a chemical solution. Your robotic device must move the chemical drum back and forth (horizontally) over a distance of 30 cm (approx. 12″) and up and down (vertically) over a distance of 13 cm (approx. 5″) at any point along the horizontal distance.
- Your solution/device must involve transmission of hydraulic (liquid) or pneumatic (air) power through the use of hypodermic syringes.
- Your solution/device must increase the mechanical advantage of the input force applied.
- The winning device is the one that requires the shortest time interval to move the drum without dropping it or spilling any of the contents.
- You must document your work in a portfolio that includes:

NOTES

☐ Information gathered from resources
☐ Sketches of all the possible solutions you considered
☐ Charts/graphs/tables showing how your device performed
☐ Illustrations/descriptions of the technology, science, and mathematics principles used in the solution
☐ Evidence of technology, science, mathematics work completed during the activity
☐ Notes made along the way

Constraints

- You may not touch the drum with any part of your body to assist in moving it.
- Your solution/device must be able to fit into a box no larger than 28,000 cm³ (one cubic foot) in volume.

Engineering Design Process

1. **Define the problem.** Write a statement that describes the problem that you are going to solve. The Design Brief and Challenge provide pertinent information.
2. **Brainstorm, research, and generate ideas.** The Research & Design section, beginning on page 3, provides background information and activities that will help with your research.
3. **Identify criteria and specify constraints.** Refer to the Criteria and Constraints listed on page 1 and 2. Add any others that your teacher recommends.
4. **Develop and propose designs and choose among alternative solutions.** Remember to save all the proposed designs so that you can turn them in later.
5. **Implement the proposed solution.** Once you have chosen a solution, determine the processes you will use to make your device. Gather the tools and materials you will need. Make sure you understand and follow all safety rules.
6. **Make a model or prototype.** The Modeling section, beginning on page 18, provides instructions for building the superstructure, linkages, and mechanisms.
7. **Evaluate the solution and its consequences.** The Evaluation section, beginning on page 19, describes how to test and analyze your device and its movements.
8. **Refine the design.** If instructed by your teacher, make changes to improve your design.
9. **Create the final design.** If instructed by your teacher, make a new device based on the revised design.
10. **Communicate the processes and results.** Be sure to include all the documentation listed under Criteria.

Research & Design

In this section, you will learn how syringes can be used to create hydraulic and pneumatic systems. See **Fig. 1-2**. Later you will be using similar hydraulic or pneumatic mechanisms to create your solution.

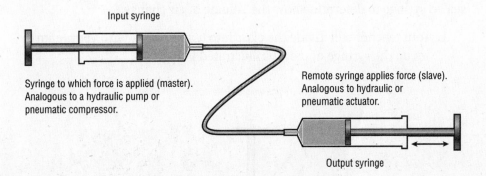

Input syringe

Syringe to which force is applied (master).
Analogous to a hydraulic pump or
pneumatic compressor.

Remote syringe applies force (slave).
Analogous to hydraulic or
pneumatic actuator.

Output syringe

Fig. 1-2 Basic hydraulic/pneumatic system using syringes

FORCES AND FLUIDS

Science   Fluid power is the use of liquids or gases under pressure to move objects or perform other tasks. Fluid-power systems may be hydraulic or pneumatic. Hydraulic systems use liquids. Pneumatic systems use gases.

Boyle's Law

The liquids or gases in a fluid-power system receive pressure from an outside source, such as a compressor. When pressure is applied to a gas, the volume of the gas decreases. The relationship between the pressure used and the volume of the gas is described in Boyle's law. **Boyle's law** states that the volume of a gas is inversely proportional to the pressure applied, so long as the temperature of the gas remains the same. This provides both advantages and disadvantages in pneumatic systems. The following experiment provides a demonstration of Boyle's law.

NOTES

MATERIALS AND EQUIPMENT

Quantity	Description
1	syringe testing apparatus
1	basin of hot water
1	syringe

For this experiment, you will place different amounts of weight on a sealed syringe to determine how the volume of air changes.

1. Your teacher will divide the class into teams. Work with your team to set up the syringe apparatus illustrated in **Fig. 1-3**.

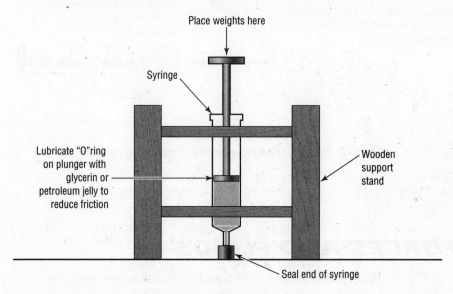

Fig. 1-3 Syringe apparatus for experimenting with the relationship between volume and pressure (Boyle's Law)

2. Fill the syringe with air and seal the tip.

3. The cubic centimeter graduations on the syringe are, for practical purposes, equivalent to milliliters (one milliliter = 1.000028 cubic centimeters). Place increasing amounts of weight on the plunger of the syringe and record the results in the table shown below.

	Volume of Air			Average Volume
Weight	**Trial 1**	**Trial 2**	**Trial 3**	

Does Boyle's law apply to liquids? Find out with this experiment.

NOTES

1. Remove the plunger and fill the syringe with water.

2. Repeat the experiment above, applying the same weights to the plunger of the syringe. Complete the table below.

Weight	Volume of Water			Average Volume
	Trial 1	Trial 2	Trial 3	

Charles' Law

Charles' law states that the pressure of a gas will change in direct relationship to temperature when volume is held constant. One principle underlying Charles' law can be illustrated with the following experiment.

1. Lubricate the bore of the syringe and the seals with a silicon lubricant.

2. Fill the syringe about halfway with air and seal the end. Record the position of the plunger using the graduations on the cylinder.

3. Immerse the syringe in hot water for about three minutes.

4. Remove the syringe from the water. Observe and record the position of the syringe plunger.

Pascal's Principle

Pascal's principle states that when force is applied to a confined liquid, the resulting pressure is transmitted unchanged to all parts of the liquid. You can see this principle in action by doing the following experiment.

MATERIALS AND EQUIPMENT

Quantity	Description
1	syringe testing apparatus
1	rule, graduated in millimeters
1	Vernier caliper, graduated in millimeters (optional)

1. With your team, set up the apparatus shown in **Fig. 1-4** on page 6. Both syringes should be 10 cc in size.

Project 1 Research & Design

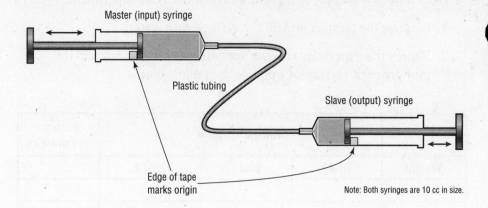

Master (input) syringe

Plastic tubing

Slave (output) syringe

Edge of tape
marks origin

Note: Both syringes are 10 cc in size.

Fig. 1-4 Experiment to demonstrate Pascal's Principle

2. Make sure that you remove all the air from the system. Why do you think all of the air needs to be removed?

_____

_____

_____

3. Fill the master syringe with water.

4. Experiment with the apparatus by pushing the plunger of the master syringe. What happens?

_____

_____

_____

_____

5. Once you have observed how the hydraulic syringe system works, take measurements to validate what you observed. Mark an origin point on each of the syringes with a small piece of masking tape as shown in Fig. 1-4. Then move the plunger in the master syringe in 10 mm increments, recording the corresponding amount the slave plunger moves in table below. Use a rule, not the graduations on the syringe. Note that there is a place on the table to enter the diameter of the plunger, but this will be done later.

NOTES

Master Syringe Size = 10 cc	Slave Syringe Size = 10 cc	Slave Syringe Size = 20 cc	Slave Syringe Size = 30 cc
Plunger dia. =	Plunger dia. =	Plunger dia. =	Plunger dia. =
Distance Moved (mm)	Distance Moved (mm)	Distance Moved (mm)	Distance Moved (mm)
10			
20			
30			
40			
50			
60			
70			
80			

6. Once you have recorded all of the data for the 10 cc slave syringe, replace it with a 20 cc syringe.

7. Repeat the experiment and fill in the data in the appropriate column of the table.

8. Finally, use a 30 cc syringe as the slave syringe and enter the data in the last column of the table.

9. Remove the plungers from each of the syringes and measure their outside diameters (or the inside diameter of the syringes). These measurements should be entered in the table above. If vernier calipers or micrometers are available, use them rather than a ruler to make these measurements.

10. Is there a relationship between the diameter of the plunger and the distance that the plunger travels? In theory, the product of the area and the distance the master plunger moves is equal to the product of the area and the distance the slave plunger moves. You can test this theory by filling in the table below, using data from the table above.

	Distance Master Plunger Moved	Area of Master Plunger End	Product of Area and Distance	Distance Slave Plunger Moved	Area of Slave Plunger End	Product of Area and Distance
	d_1	a_1	$d_1 \times a_1$	d_2	a_2	$d_2 \times a_2$
10 cc slave	80 mm					
20 cc slave	80 mm					
30 cc slave	80 mm					

NOTES

THINKING CRITICALLY

1. If the contents from a can of spray paint were removed and put into a larger pressurized container, what would be the disadvantage?

_____

_____

_____

2. How much pressure could you produce in a syringe by pressing the plunger with your thumb?

_____

_____

3. How could you adapt a syringe to inflate bicycle tires?

_____

_____

_____

_____

4. What type of pressure does a barometer measure? How might changes in this pressure affect the performance of your robot?

_____

_____

_____

5. Suppose you filled a syringe with air. What might happen to it if you left it overnight, and when you came back, the temperature was ten degrees warmer?

_____

_____

_____

_____

MECHANICAL ADVANTAGE, WORK, AND EFFICIENCY

Energy is the capacity to do work or to make an effort. **Work** is using force to act on an object in order to move that object in the same direction as the force. One of the factors to be considered is efficiency. **Efficiency** is the ability to achieve a desired result with as little wasted energy and effort as possible. An efficient machine accomplishes a lot of work for the amount of energy used. Some machines are more helpful and efficient than others. The measure by which a machine increases force is known as its mechanical advantage (MA).

MATERIALS AND EQUIPMENT

Quantity	Description
1	syringe testing apparatus

Mechanical advantage allows us to produce an output force that is greater than the input force. (Force = pressure × area.) Examples include pulling nails with a claw hammer, lifting a heavy rock with a plank or steel bar, and driving screws with a screwdriver. Mechanical advantage can be calculated by the formula:

$$MA = \frac{F_r}{F_e}$$

In this equation, F_r stands for resistance force (load). This is the force applied by the machine. For example, a pry bar exerts a force on a closed lid on a box. F_e stands for effort force, the force applied to the machine. An example is the force you apply to the pry bar. If a small effort force results in a large resistance force, the machine has a high mechanical advantage. The greater the mechanical advantage, the more helpful the machine.

For every machine that gives us a mechanical advantage, there is a trade-off in the distance that the effort to the machine must be moved compared to the distance the load moves. For example, to jack up a car, we must move the lever a relatively great distance up and down, though the car moves a very small amount. We can calculate velocity ratio for machines, which is the distance the load moves divided by the distance the effort moves.

1. Your teacher will provide an apparatus like the one shown in **Fig. 1-5** on page 10. The input force is applied to the plunger of the master syringe, and the output force is delivered by the plunger of the slave syringe. The work input to the plunger of the master syringe is theoretically equal to the work output of the plunger of the slave syringe.

Work In = Work Out

Since Work = Force × Distance, the following formula is also true:

Force In × Distance In = Force Out × Distance Out

NOTES

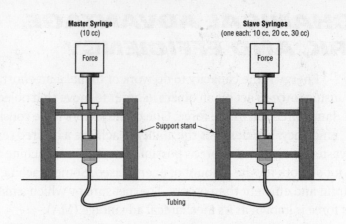

Fig. 1-5 Apparatus for determining mechanical advantage

2. Use 10 cc. syringes as the master and the slave syringes. Press the master syringe down a distance of 1 unit. Complete the following table. What can you conclude about the distance the slave cylinder moves compared to how far the master cylinder moves?

Distance Master Cylinder Moves	Distance Slave Cylinder Moves
1 unit	
2 units	
3 units	
4 units	
5 units	

_____

_____

3. The mechanical advantage of this closed hydraulic system can be determined by dividing the force applied to the slave syringe at balance by the force applied to the master syringe at balance: Place equal weights on the end of each syringe plunger. What happens?

_____

_____

Mechanical Advantage = Output Force / Input Force

What mechanical advantage did you produce in Step 2?

_____

_____

Mechanical Advantage = Input Distance/Output Distance

4. As noted earlier, there is a trade-off in distance that occurs as mechanical advantage is increased. For example, in pulling out a nail with a claw hammer, the force on the nail is increased. However, the nail is moved through a short distance relative to the distance through which the handle of the hammer moves. In a hydraulic system, pushing on the plunger of a smaller-diameter syringe increases the output force of a larger syringe, but the smaller plunger must be moved through a greater distance.

   You can also determine the mechanical advantage using the Input Distance (effort) and the Output Distance (load). Use a 10 cc syringe as the master syringe and a 20 cc syringe as the slave syringe. Measure the distance the slave cylinder moves for each distance the master cylinder moves and record these distances in the table below. Then calculate the mechanical advantage for each distance by dividing the input distance by the output distance.

Input Distance (master cylinder, 10 cc)	Output Distance (slave cylinder, 20 cc)	Mechanical Advantage (Input Dist./Output Dist.)
1 unit		
2 units		
3 units		
4 units		
5 units		

_____

_____

_____

5. You may have noticed that in this experiment that the plunger of the master cylinder requires considerable effort to move it, even though there is nothing attached to the slave cylinder. This means that the work output is less than the work input. This makes the system less inefficient because some energy has been wasted. If so, where do you think the loss of energy occurred?

_____

_____

_____

The actual efficiency of the syringe system could be determined through the application of the following relationship:

Efficiency = (Work Out/Work In) × 100

NOTES

THINKING CRITICALLY

1. If a force exerted on a master piston causes a slave piston to move five times as far, how does the force on the slave piston compare with the force on the master piston? What is the mechanical advantage?

_____

_____

2. How could the friction created by the plunger seals be reduced?

_____

_____

3. When mechanical advantage is increased, what are the tradeoffs?

_____

_____

_____

4. For what type of system might a person design a hydraulic cylinder system that has a mechanical advantage of less than 1?

_____

_____

_____

FORCE, PRESSURE, AND DISTANCE

MATERIALS AND EQUIPMENT

Quantity	Description
1	syringe testing apparatus shown in Fig. 1-5

1. In your earlier experiments, you discovered that Force = Area × Pressure and, therefore, Pressure = Force/Area. These relationships can be graphed as shown in **Fig. 1-6**. What relationships exist between Force and Area if Pressure remain constant?

_____

_____

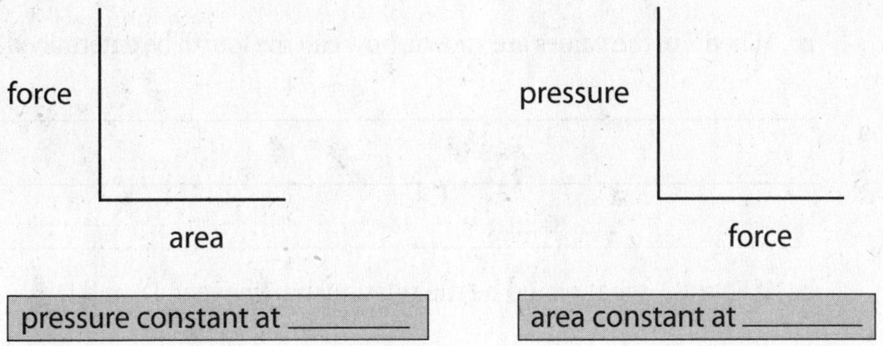

force |_____

area

pressure |_____

force

| pressure constant at _____ |

| area constant at _____ |

Fig. 1-6 Force equals area times pressure.

2. Since the syringe system with which you are working is closed, then according to Pascal's principle, the pressure is constant throughout the system. If Force = Area × Pressure for each syringe, develop a proportion that is true for the system.

_____

_____

3. Suppose two syringes in the closed system have plungers with a face area of 2 cm^2 and 5 cm^2. If a force of 40 newtons is applied to the plunger of the smaller syringe, what pressure is present in the system? What will be the output force of the larger syringe?

_____

_____

4. If the radius of the master syringe is 2.5 cm, and the force applied to it is doubled to 80 newtons, what happens to the pressure in the system as compared to before the force was doubled?

_____

_____

5. What will happen if the master syringe is replaced with one that is larger in diameter than the original master syringe and the same force is applied?

_____

_____

NOTES

6. The concept that work in = work out was represented as $F_1D_1 = F_2D_2$ or Force In × Distance In = Force Out × Distance Out:

   a. When 3 of the values are known, how can the fourth be determined?

   _____

   _____

   _____

   b. If $F_1 > F_2$, what would be the relationship between D_1 and D_2?

   _____

   c. If $D_1 > D_2$, what would be the relationship between F_1 and F_2?

   _____

7. Suppose a force of 40 newtons is applied to the master syringe, which has a radius of 3 cm:
   a. What is the pressure in the system?

   _____

   _____

   b. If the slave syringe has a 4 cm radius, what is the output force?

   _____

   c. If the master syringe moved is 1.8 cm, what is the distance that the slave syringe plunger moved?

   _____

   _____

8. Recall that $F_1 = A_1P$ and $F_1 = A_2P$ and that the pressure is constant throughout a closed system. Develop a mathematical model that relates the distance a plunger travels, D, with area of plunger face, A.

   _____

System Relationships

When you build your robotic device, the equipment will have limitations regarding size. Each syringe you select will have a maximum distance that it is capable of delivering. You must consider this when you design your solution. You will want to try several combinations of syringe systems to find out which ones work best.

For example, suppose the drum must be lifted 13 cm. If your robot design uses only one syringe (slave) to lift the arm and it is of sufficient possible range, what sizes of master syringes would be workable in order to achieve the lift of 13 cm?

Name________________________________ Date________________ Class ________________

NOTES

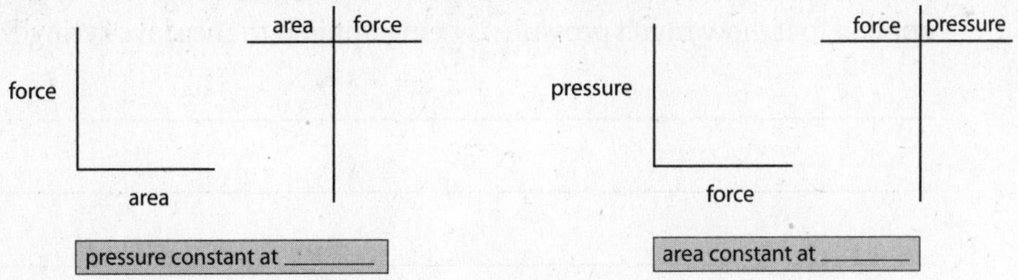
area
force
force
area
pressure constant at ________
force
pressure
pressure
force
area constant at ________

Fig. 1-7 T-graphs help determine which relationships work best.

If only shorter syringes are available for the master syringe, how should its radius compare to the radius of the slave syringe? Suppose only syringes longer than needed are available. How should its radius compare to the radius of the slave syringe?

Construct T-tables and graphs of your data to help you determine which relationships work best. See **Fig. 1-7**. When you experiment with force, use this table to record your data:

Force$_1$ in newtons	Distance$_1$ in mm	Force$_2$ in newtons	Distance$_2$ in mm

Explore various ways of setting up proportions to solve problems. For example:

$$\frac{F_1}{D_1} \quad \frac{F_2}{D_1} = \frac{F_1}{F_2} \quad \frac{D_1}{D_2}$$

THINKING CRITICALLY

1. If the slave syringe moves a distance of 1.5 cm and lifts a weight of 3,000 gm when you move the master syringe 3 cm, what force would you be applying to the plunger of the master syringe?

____

____

____

Name_____ Date_____ Class _____

2. If the master syringe has a radius of 0.80 cm, and a 2000 gm force is applied to it, how much pressure is being applied to the slave syringe?

_____

_____

_____

DESIGN YOUR SOLUTION

You will now apply the basic operating principles of hydraulic and pneumatic systems to the design of your solution. One student's design is shown in **Fig. 1-8**. Your teacher will demonstrate some systems that may give you additional ideas.

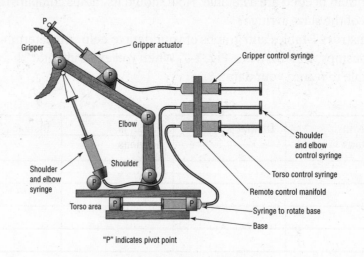

Fig. 1-8 One student's solution

MATERIALS AND EQUIPMENT

Quantity	Description
2	20 cc syringes
2	30 cc syringes
2	25 cm lengths of clear plastic tubing
1	35 mm plastic film container
1	platform representing the loading dock

The 35 mm film container represents a chemical drum. The film container is 32 mm (approx. 1.25″) in diameter and 51 mm (2″) in height. Fill it with water to within 3 mm (⅛″) of the top. The water represents the chemical solution. Try systems that consist of one 20 cc syringe connected to one 30 cc syringe by a 25 cm length of tubing. Your system could be pneumatic (filled with air) or hydraulic (filled with water).

As you develop your design, list the advantages and disadvantages of each type of system. For example, air is compressible. Thus, there is no positive displacement of the slave syringe when the master syringe is moved. There is positive displacement with the hydraulic system, but the water often leaks. In addition, it is somewhat difficult to remove the air bubbles from the water.

Keep in mind that the diameters of a given pair of syringes are fixed parameters. The displacement (distance moved) of one piston (plunger) in a hydraulic system and the corresponding displacement of the other are variables and these variables are directly related.

1. Connect two different-sized syringes together with the tubing and operate them as a pneumatic system. Record your observations.

2. Then fill the tubing and syringes with water. Observe the force required to move the plunger on the larger syringe by pushing the smaller one and vice-versa. Observe the distances that the plungers move relative to one another. Record your observations.

3. With materials provided by your teacher, try various methods of fastening syringes to other components and linkages. Brainstorm ways in which the body of a syringe can be mounted so that it is not distorted or damaged.

4. How can you change one form of motion into another using the syringes? For example, can you change the linear movement of the syringe to rotational movement? Rotational movement might be desirable for the base of a robot, for example.

5. Prepare sketches or drawings of at least one solution to the Challenge.

THINKING CRITICALLY

1. Observe some of the designs being worked on by your classmates. Which designs do you think will work best? Why?

   _____

   _____

2. For what kinds of jobs do you think hydraulic robots would be preferred to pneumatic robots and vice versa?

   _____

   _____

   _____

NOTES

Modeling

For this part of the project, you will use tools and equipment to build the superstructure, linkages, and mechanisms for your robotic device.

SAFETY FIRST

Before You Begin Make sure you understand how to use the tools and materials safely. Ask your teacher to demonstrate their proper use. Follow all safety rules. Refer to the "Safety Handbook," beginning on page 291, for more information about safety in the lab.

BUILD YOUR ROBOTIC DEVICE

Engineering Technology

Think about all the things you have learned about pneumatic and hydraulic systems. Use that information to help you engineer your robot.

MATERIALS AND EQUIPMENT

Quantity	Description
6 to 8	syringes in assorted sizes (e.g., 10 cc, 20 cc, 30 cc)
120 cm	clear plastic tubing
1	35 mm plastic film container
Assorted	materials, fasteners, and adhesives

1. Using your drawings and the data you have gathered, build your robotic device and test it.

2. Re-engineer your design, as needed.

3. Check to be sure your design meets all the criteria and constraints.

THINKING CRITICALLY

1. If you had to lift a car using a hydraulic jack, would you want the cylinder that lifts the car to be many times larger or smaller in area than the cylinder to which the jack handle is connected?

_____

_____

_____

Project 1 Modeling

Evaluation

During this part of the project, you will test your solution to see how well it meets the criteria and constraints specified in the Design Brief. Your teacher will time each test. The fastest device will win the Challenge. Then you will analyze the movements made by the robot.

TEST YOUR ROBOT

Your teacher will place the chemical drum and loading dock in position, as shown in **Fig. 1-9**. You may place your robotic device wherever you wish during the test.

MATERIALS AND EQUIPMENT

Quantity	Description
1	loading dock (block of wood, 13 cm thick)
1	plastic 35 mm film container

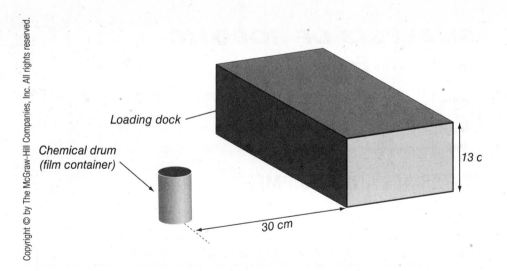

Loading dock

Chemical drum
(film container)

13 c

30 cm

Fig. 1-9 Layout for the challenge

Project 1 Evaluation

NOTES

1. When your turn comes, place your robotic device in position.

2. When instructed to do so, use your robot to pick up the chemical drum and place it on the loading dock. Be careful not to spill the drum's contents.

3. Record your time.

THINKING CRITICALLY

1. If you were to redesign your robot, how would you do it?

   _____

   _____

   _____

   _____

2. Was your robot powered pneumatically or hydraulically? If you used the alternate system, how do you think your test results would have changed? Why?

   _____

   _____

   _____

   _____

ANALYSIS OF ROBOTIC MOVEMENT

Math

During this analysis, you will determine how location of the drum and the robotic arm affected the results.

MATERIALS AND EQUIPMENT

Quantity	Description
1	loading dock (block of wood, 13 cm thick)
1	plastic 35 mm film container
1	volunteer robot (similar to the one shown in Fig.1-8)

Project 1 Evaluation

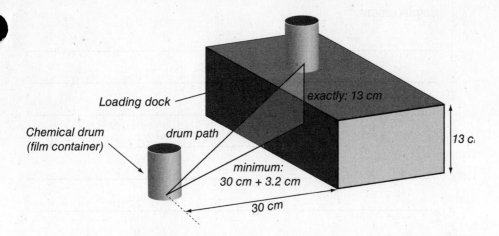

Loading dock

Chemical drum
(film container)

drum path

exactly: 13 cm

13 c.

minimum:
30 cm + 3.2 cm

30 cm

Fig. 1-10 The path the drum follows can be shown as a triangle.

Placement of the robot and drum during the test determined the range of motion your robot needed. A triangle can be used to show how the arm needs to move in order for the drum to be lifted 13 cm. See **Fig. 1-10**. One side of the triangle represents the initial position of the arm as it makes contact with the drum. Another side represents the arm after it has lifted the drum 13 cm. The third side represents the length of rise of the drum. (*Note:* Since the drum is on the end of the arm that is rotating upward, the drum path actually follows an arc, but a triangle works for this purpose.) Figure out a range of positions at which you could place the drum and still have the volunteer robot pick it up from the loading dock.

Using the volunteer robot, gather data for the following questions:

1. What does this upward motion mean in terms of the distance that the slave syringe's plunger must move?

_____

_____

_____

_____

2. What should be accounted for in the master syringe?

_____

_____

_____

_____

Project 1 Evaluation

NOTES

3. Isolate the horizontal movement. Determine the horizontal displacement.

_____

_____

_____

_____

_____

_____

Does the arm need to be located at any particular point between the drum and platform in order to make this horizontal transfer? Try placing the arm at different locations between the drum and platform.

4. If the arm cannot rotate and the platform is fixed, where does the drum need to be located?

_____

_____

5. Using the Pythagorean theorem ($a2 + b2 = c2$), estimate the length of the drum path. Knowing that the actual drum path is an arc, is this estimate a little under or a little over?

_____

THINKING CRITICALLY

1. Why did the drum need to be raised exactly 13 cm?

_____

_____

2. Why did the drum have to be moved forward 30 cm plus the diameter of the drum?

_____

_____

_____

_____

Composite Beam

Design Brief

The best construction materials are strong, yet lightweight. Wood is surprisingly strong for its weight and is, therefore, well suited for building houses. Steel-reinforced concrete beams, rather than wood, are often used in the construction of larger buildings. See **Fig. 2-1**. However, concrete and steel are both heavy, and that presents some problems in construction. A lighter weight but equally strong alternative to steel-reinforced concrete would be a best-seller in the construction industry. One alternative might be to reinforce the concrete beam with a material other than steel—ideally, a recycled material.

Fig. 2-1 Construction of a large building using steel-reinforced concrete beams

Challenge

Design the lightest-weight and strongest beam by reinforcing concrete with one or more of the following materials: discarded aluminum cans, HDPE plastic containers, PETE plastic containers, and/or newspaper.

Criteria

- All beams will be cast on the same day to ensure that the beams are produced from consistent materials.
- Once cast, the beams will be set aside for curing at least 14 days.
- Beams will be weighed and tested to determine the maximum load they held at the point of failure.
- Each team will determine the efficiency of its beam. The efficiency is the ratio of how much weight the beam held (the load) to the total weight of the beam.
- The winning design will be the most efficient beam (the one having the highest load-to-weight ratio).

NOTES

- All designs must be accompanied by appropriate documentation, including:
  - ☐ Drawings of all possible solutions your team considered
  - ☐ Drawing of the mold design for the beam
  - ☐ Stress Analysis Table showing how materials reacted to stress
  - ☐ Beam Performance Table showing how your team's beam performed when tested to the point of failure
  - ☐ Your team's report hypothesizing why your team's beam failed
  - ☐ Information gathered from resources
  - ☐ Notes made along the way

Constraints

- The beam must be cast in a reusable mold or reusable form that you design and construct.
- The beam must be 40 cm (about 16″) long and fit within a volume of 1050 cubic centimeters (about 64 cubic inches).
- It must be made from concrete and one or more of the materials identified in the Challenge.

Engineering Design Process

1. **Define the problem.** Write a statement that describes the problem. The Design Brief and Challenge provide information.
2. **Brainstorm, research, and generate ideas.** The Research & Design section, beginning on page 25, provides background information and activities that will help with your research.
3. **Identify criteria and specify constraints.** Refer to the Criteria and Constraints listed on pages 23 and 24. Add others your teacher recommends.
4. **Develop and propose designs and choose among alternative solutions.** Save all proposed designs so that you can turn them in later.
5. **Implement the proposed solution.** Once you have chosen a solution, determine the processes you will use to make your beam. Gather the tools and materials you will need. Make sure you understand and follow all safety rules.
6. **Make a model or prototype.** The Modeling section, beginning on page 38, provides instructions for building the form, casting the beam, and checking dimensions.
7. **Evaluate the solution and its consequences.** The Evaluation section on page 48, describes how to test and analyze the beam.
8. **Refine the design.** If instructed by your teacher, make changes to improve your design.
9. **Create the final design.** If instructed by your teacher, make a new beam based on the revised design.
10. **Communicate the processes and results.** Be sure to include all the documentation listed under Criteria.

Research & Design

This section provides information about the characteristics and properties of materials. You can learn about the types of stress to which materials and structures are subjected and conduct experiments on the stress characteristics of the materials you will be using to make your composite beam. This information can help your team as you prepare possible designs and then decide which design is most likely to meet the criteria and constraints.

RECYCLING

The symbols in **Fig. 2-2** are found on two common types of recyclable plastics. The plastics are polyethylene terephthalate (PETE) and high density polyethylene (HDPE). Look for these symbols when choosing plastic materials to reinforce your concrete beam. Plastic soda bottles are made from PETE, and milk jugs are made from HDPE. What other containers can you find that are made from PETE or HDPE?

Polyethylene Terephthalate (PETE) Plastic

Industrial Recycling

Once a material is turned in for recycling, it is often remade into the same product. This is what is known as **closed-loop recycling**. For example, aluminum cans are melted down in order to be made into more aluminum cans. Newspaper is shredded and turned into pulp from which more newsprint can be made. Thus, the closed loop consists of making the product, using the product, and then recycling it to be made back into the same product again.

Though some materials can be recycled into the same product, other materials must be recycled into different products. For example, PETE plastic soda bottles cannot readily be recycled into new soda bottles. Instead, they can be processed into fibers that are for insulation or clothing. This type of recycling is known as **conversion recycling**.

High Density Polyethylene (HDPE) Plastic

Fig. 2-2 Look for either (or both) of these two symbols when choosing the plastic materials for your composite beam.

NOTES

THINKING CRITICALLY

1. What other materials besides plastic can be recycled in a closed loop?

_____

_____

2. Name some recyclable materials other than those already identified in class that must be converted to another form in order to be useful. Explain how the new form is used.

_____

_____

_____

3. What are some of the advantages and disadvantages to the use of plastic versus glass milk bottles?

_____

_____

_____

_____

_____

CLASSIFICATION OF MATERIALS

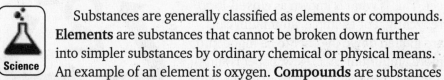

Substances are generally classified as elements or compounds. **Elements** are substances that cannot be broken down further into simpler substances by ordinary chemical or physical means. An example of an element is oxygen. **Compounds** are substances whose smallest units are made up of two or more elements that are chemically combined. Sodium chloride (table salt) is an example of a compound. It is made up of the elements sodium and chloride.

A **mixture** is two or more substances that can be separated by physical means. The substances could be elements or compounds, but they are not chemically bonded. Sand and gravel mixed in the same container is an example of a mixture.

Mixtures in which the substances can be distinguished visually are called **heterogeneous mixtures**. The mixture of sand and gravel is an example of a heterogeneous mixture.

Homogeneous Mixtures

In a **homogeneous mixture**, the substances cannot be visually distinguished from one another.

A **composite** is a mixture that results in a new material with properties that are more desirable than either of the original materials by themselves. The sand mix from which you will build your beam is a composite consisting of sand and portland cement. Concrete is a composite of sand, gravel, and cement. Tennis rackets are often composites of graphite fiber and plastic resin. Boats are often made of a composite of glass fiber and plastic resin. By themselves, none of the materials that were combined to make these composites has the necessary properties to serve alone in the applications listed. However, when the materials are combined to make the new material—the composite—they serve the applications very well.

THINKING CRITICALLY

1. Describe a method that could be used to separate a mixture of sand and iron filings.

   _____

   _____

2. When you use concrete and recyclable materials to make your composite beam, is the resulting material a compound or a mixture? Why

   _____

   _____

   _____

   _____

3. In general, would compounds or mixtures have more resistance to being pulled apart? Why?

   _____

   _____

   _____

   _____

   _____

4. Use a microscope to observe the materials you will use in the composite beam. The materials are listed below. Samples of other materials could also be included among those you observe. Sketch and verbally describe what you observe.

NOTES

STRESS AND STRAIN

Math

Four forces that can act on structural materials and components are shown in **Fig. 2-3**. **Tension** pulls materials apart. **Compression** squeezes materials together. **Torsion** is a twisting force. **Shear** tends to make portions of a material or structure move in parallel but opposite directions. These outside forces can create forces within structural materials and/or components. These inner forces are called **stress**. The stress can change the original size or shape of the materials and/or components. This resulting change in size or shape is called **strain**. The stress and resulting strain can jeopardize the integrity (soundness) of the material and/or the structure. It is very important to understand these forces of tension, compression, torsion, and shear when designing a structure and its components, and when choosing construction

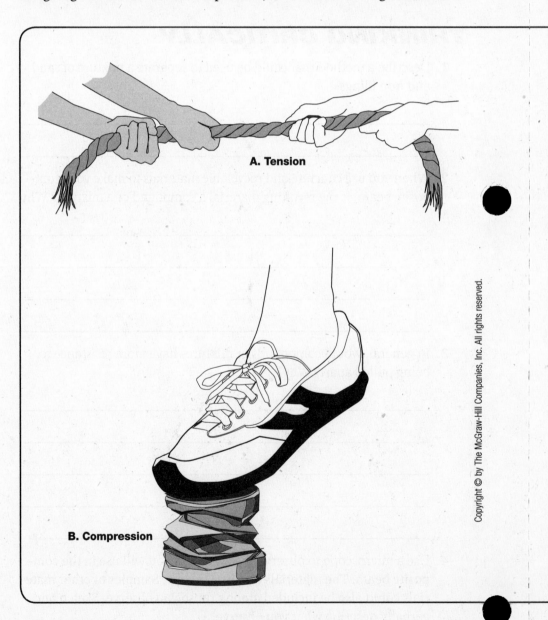

A. Tension

B. Compression

NOTES

materials in order to minimize the stresses that these forces can put upon the structure.

You can see forces in action with a simple experiment. Obtain a wide rubber band and stretch it. By stretching the rubber band, you are putting it in tension. The tensioned rubber band has a different shape than it did originally. This is an example of strain. Note the direction in which you are applying the force.

Apply force to a gum rubber eraser by holding the ends of the eraser between your fingertips and then squeezing your fingers toward each other. Observe that the sides of the eraser tend to squeeze out or deform in relation to the force you are applying to the ends of the eraser. You are putting stress on the eraser and, as a result, the eraser is in compression. The stress results in a change in shape. This is another example of strain.

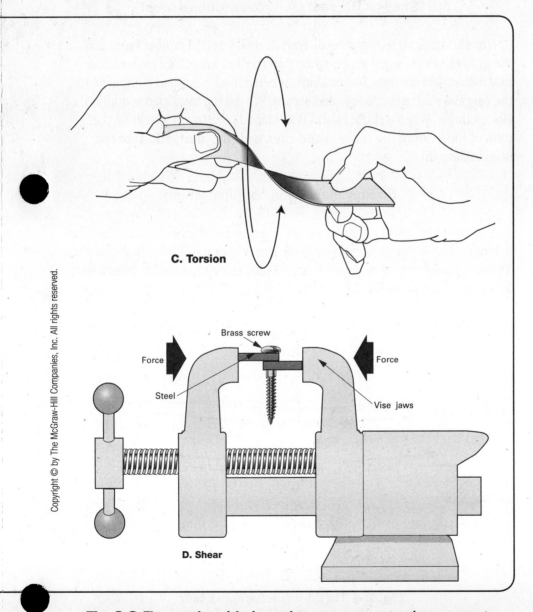

C. Torsion

D. Shear

Fig. 2-3 These are four of the forces that can act on construction components.

NOTES

Note that when a material is in compression, the forces are directed toward one another. Compare this to tension, in which the forces are directed away from one another.

Figure 2-4 shows a beam supported at its ends. A load is pressing down on the beam. The upper part of the beam undergoes compression and tends to be pushed together, while the lower part of the beam undergoes tension and tends to be pulled apart. The middle part of the beam is considered a "dead area." This is why holes are cut in the middle of structural members of an aircraft, reducing weight while minimizing reduction in strength. It is also why holes in floor joists for the routing of wires and pipes should be drilled in the middle of these structural members.

In engineering terms, stress is the ratio of the applied load to the original cross-sectional area of the material:

$$\text{Stress} = \text{F (Load)} / \text{A (Cross-sectional area)}$$

Stress is difficult to measure directly. Recall how the rubber band and the gum eraser changed shape when you applied force. The ratio of how much an object changes size or shape when a load is applied compared to the original size or shape is called **strain**. When a tension load is applied, (for example, when a rubber band is stretched) the strain would be the ratio of the amount the rubber band stretched (the **elongation**) to the original length:

$$\text{Strain} = \text{elongation} / \text{original length}$$

For example, assume that an unstretched piece of rubber band was 5 cm in length, but when a load was applied, the rubber band stretched to 8 cm. The elongation would be 8 cm–5 cm = 3 cm. Since the original length was 5 cm, the strain would be 3 cm / 5 cm = 0.6.

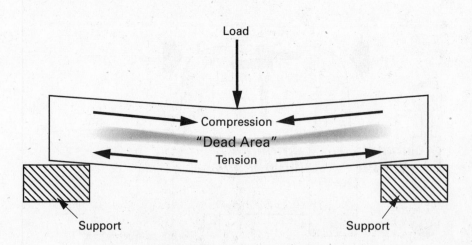

Fig. 2-4 The stresses acting upon a beam

When materials are tensioned (stretched), there is a range in which they will return to their original shape when the force is removed. (This ability of a material to return to its original shape once a deforming load is removed is called **elasticity**.) The strength at the upper limit of this range is called the **yield strength**. **See Fig. 2-5**. After the yield point has been exceeded, the material will be permanently deformed. Eventually, as the tension stress continues to increase, the material will break. The maximum stress at the point when the material failed is called the **ultimate strength**. When the stress is below the yield strength limit (where there is no permanent deformation of the material), there is a direct relationship between stress and strain. Stress can be expressed as a function of strain with the following equation:

$$Stress = constant \times strain$$

The constant in the above equation is called the *constant* (or *modulus*) *of elasticity* or *Young's modulus*.

Knowledge of stress and strain can help us determine whether or not a material is suitable for a particular application. People often check the stress capabilities of materials without really thinking about it. For example, a person might pull on a rope to determine if it is sufficiently strong to support the weight of his or her body, or a person may try to push on a board to determine if it is strong enough to support his or her weight.

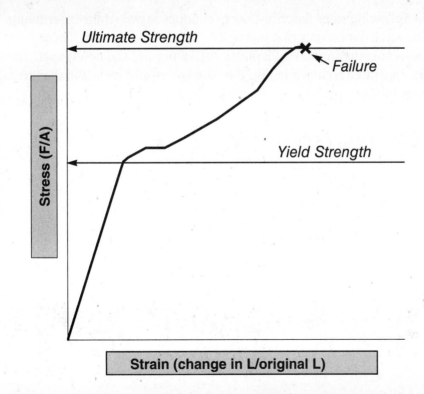

Fig. 2-5 A Stress/strain curve for a sample material

Project 2 Research & Design

NOTES

Knowing that stress is directly related to strain and that strain is indicated by the amount a material elongates or otherwise changes shape as the result of an applied force allows us to test and evaluate materials before applying them to a situation.

Although the constant of elasticity of a solid is approximately the same for tension and compression, the ultimate strength may be very different for the two cases. Concrete, for example, is very strong in **compressive strength** (ability to resist squeezing forces). That is why it is used in such applications as building foundations, where it is subjected to the heavy weight of the building. However, concrete is weak in **tensile strength** (the ability to withstand stretching forces). This is why materials with high tensile strength are used to reinforce concrete. Typical materials used for this purpose include metal mesh, wires, and bars.

Analyzing Stress and Strain
MATERIALS AND EQUIPMENT

Quantity	Description
assorted	HDPE plastic (such as milk jugs); PETE plastic (such as soda bottles); aluminum beverage cans; newspapers
as needed	scissors
as needed	tin snips

The following steps describe how to conduct stress/strain experiments on the recycled materials that you will be using in your beam. These materials are aluminum, HDPE plastic, PETE plastic, and newspaper. The testing apparatus is shown in **Fig. 2-6**. You will record results in the Stress Analysis Table.

STRESS ANALYSIS TABLE

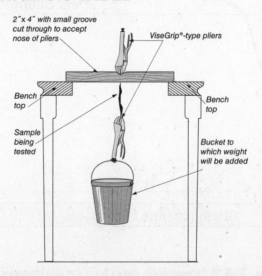

Fig. 2-6 Apparatus for testing the stress and strain of the recycled materials

Elongation in Millimeters				
Weight in Bucket	Aluminum	PETE Plastic	HDPE Plastic	Newspaper
200 g				
400 g				
600 g				
800 g				

1. Use scissors and tin snips to cut the materials into strips that are 0.5 cm wide and 30 cm long.

2. Measure the thickness of each material using a micrometer or vernier caliper to determine which material is thickest. Strips of the thinner materials should be layered so that each testing sample is approximately the same thickness.

3. To place a sample in the testing apparatus, pass it through the groove cut in the 2 × 4 and tighten the ViseGrip® pliers securely to the bottom and top edges of the sample.

4. Add water or sand to the bucket to achieve the first weight (200 g) on the Stress Analysis Table. Fasten the bucket to the lower pliers.

5. Measure the length of the material while it is under the 200 g tension. Subtract the original 30 cm length from this new length. The difference is the elongation. Record the elongation on the Stress Analysis Table.

6. Test each sample using the 200 g weight. Record the elongation for each sample.

7. Add more sand or more water to the bucket to achieve 400 g.

8. Test each sample at this weight, measuring each one and subtracting the length under this new tension from the original 30 cm length to determine elongation. Record results on the Stress Analysis Table.

9. Continue testing each sample at each weight and recording the elongation. When a sample fails, write the word "failed" in the box for the weight at which the material could no longer withstand the stress.

Graphing Data

The experiments that you performed resulted in data on how much PETE, HDPE, aluminum, and newspaper elongated as the load applied to them increased. Data about strength properties of materials is useful in deciding which materials to select for a product and how to use them. For example, the data you recorded can help you make decisions about how you might best use the materials in the design of your beam.

NOTES

Putting data into graph form can make it easier to understand. Look at **Fig. 2-7**. What does it tell you about the relationship of stress (internal force) to strain (change in shape)?

1. Make a line graph using the data you obtained from testing the various materials. First choose an appropriate scale and then plot the data that you recorded on the Stress Analysis Table. Draw your graph on a large sheet of newsprint, or use a computer to generate the graph for projection. Make sure you label the axes on the graph.

2. Your team should present its stress/strain graph to the class. As a class, compare graphs of different materials and describe the points in common among the graphs. Discuss the differences between the groups. What makes it easier or more difficult to compare graphs?

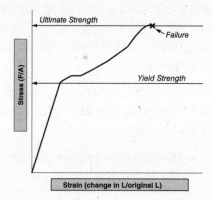

Fig. 2-7 The relationship between stress and strain can be shown graphically

THINKING CRITICALLY

1. If a stereo speaker is to be hung from a steel rod, how can one reduce the stress applied to the rod?

_____

_____

_____

2. When a weight is hung from your composite beam, what types of stresses will act on it?

_____

_____

3 Look at the two graphs in **Fig. 2-8**. How can you determine which company has greater sales?

_____

_____

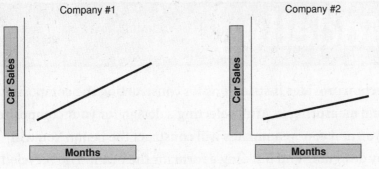

Fig. 2-8 Graphs showing sales of two car companies

4. Look at the graph in **Fig. 2-9**. Suppose that you have invented a re-markably strong new material that has potential for beam construc-tion. The graph depicts the strength of this material when subjected to increasing force. How might you rescale the graph in the figure so that the effect is more dramatic?

_____

_____

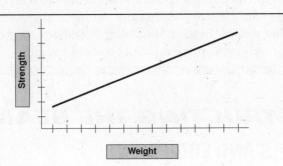

Fig. 2-9 Strength of material with increasing load

Designing the Beam

You have learned about the properties of materials, so you are ready to design your beam. Two possible solutions are in Fig. 2-9 on page 39.

1. With your team, brainstorm at least three different designs.

2. Make sketches of each design and discuss the pros and cons with your team. Save all sketches so that you can turn them in later.

3. With your team, select one design to implement.

4. Prepare three-view drawings of the chosen design.

THINKING CRITICALLY

1. Which materials did you choose to reinforce or lighten your beam?

_____

2. Aluminum has good tensile strength but a low coefficient of friction. How might that affect the strength of a beam?

_____

_____

NOTES

Modeling

This section provides instructions for constructing the composite beam and measuring it. After selecting a design for your composite beam, you and your teammates will construct the beam. You will begin by designing and building a form for the beam. The recycled materials you have decided to use to reinforce the beam and/or to lighten its weight will be placed in the form. Next, sand mix (sand and cement) will be poured into the form. This process is called "casting." After the beam has hardened, you will remove it from the form and measure it.

SAFETY FIRST

Before You Begin Make sure you understand how to use the tools and materials safely. Have your teacher demonstrate their proper use. Follow all safety rules. Refer to the Safety Handbook, beginning on page 291, for more information about safety in the lab.

CONSTRUCTING THE BEAM

MATERIALS AND EQUIPMENT

Quantity	Description
2 sq. ft	exterior plywood, ½" to ¾" thick (for bottom, sides, and ends of form)
5 lbs.	sand mix (dry, sacked mixture of sand and cement)
as needed	water (as per instructions on sand mix sack)
1	plastic mortar pan or similar container for mixing sand mix
1	mortar trowel or similar tool to mix and transport sand mix
1 sheet	9' × 12' polyethylene plastic sheet, 4-6 mil thick
assorted	HDPE plastic (such as milk jugs); PETE plastic (such as soda bottles); aluminum beverage cans, newspapers
assorted	hand and/or power tools for cutting recycled materials and building the forms

SAFETY FIRST

Before You Begin Make sure you understand how to use the tools and materials safely.

Designing and Building the Form

Engineering Technology

     After your team has agreed on a single beam design, each team member should prepare a three-view drawing of what he or she thinks would be a good design for the form (mold) for that beam. A pictorial drawing of a sample design is shown in **Fig. 2-10**. Team members should then evaluate each other's drawings and decide which drawing should be used when building the form. Once a drawing has been selected, team members should work together to build their form. Certain beam shapes will require draft so that the form pieces can be separated from the beam. Refer to **Fig. 2-11**.

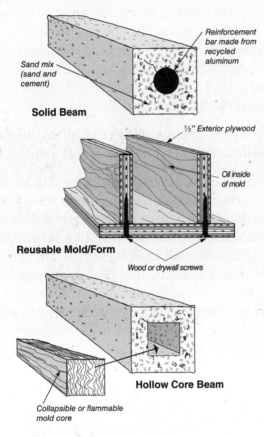

Reinforcement bar made from recycled aluminum

Sand mix (sand and cement)

Solid Beam

½" Exterior plywood

Oil inside of mold

Reusable Mold/Form

Wood or drywall screws

Hollow Core Beam

Collapsible or flammable mold core

Fig. 2-10 Sample solutions

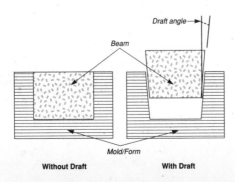

Draft angle

Beam

Mold/Form

Without Draft   **With Draft**

Fig. 2-11 Draft angle allows the mold to be easily released from the beam

Project 2 Modeling

NOTES

Casting the Beam

1. Prepare the recycled materials. You may need to cut or shred the plastic and/or newspaper and crush the aluminum cans.

2. Before you place the recycled materials into the form for your beam, weigh each of them. Record the data on the Beam Performance Table. This table will also be used later to record the weight of the beam immediately before testing as well as the maximum load that it held.

3. Place the recycled materials into the form.

It is important that all the beams be poured on the same day and from the same batch of sand mix so that the sand mix is homogeneous for all teams and so that the curing times and conditions will be identical. Therefore your teacher will designate a specific day on which the sand mix will be placed in the form. Your teacher will prepare the sand mix.

4. Add the prepared sand mix to the form.

5. Place the filled form away from sunlight or other sources of intense heat so moisture does not exit too quickly during curing time.

Concrete does not harden by drying. It hardens by means of a chemical process called "curing." When the ingredients are combined with water, a chemical reaction occurs, causing the concrete to harden, or *cure.* If water is no longer available, then the reaction stops permanently. Consequently, the strength of concrete will never exceed the strength that the concrete had when the moisture was used up. This is why concrete sidewalks are covered with plastic, straw, or other materials. The covering prevents the evaporation of moisture so that the concrete can realize its ultimate strength. Polymers may be added to reduce evaporation from the surface.

The strength of concrete is affected by the curing method and the curing time. The graph in **Fig. 2-12** shows the relationship of curing time to the ultimate strength of concrete.

The curing of concrete is an exothermic (heat-producing) reaction. You can observe the heat by holding your hand on the beam after it bigins to cure.

Project 2 Modeling

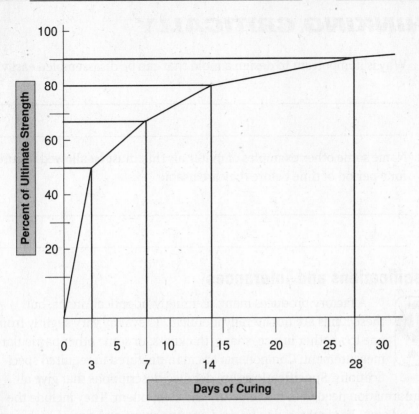

Fig. 2-12 Ultimate strength of concrete in relation to curing time

BEAM PERFORMANCE TABLE

Reinforcement and Weight Reduction				Testing Results		
Weight of Aluminum in grams	Weight of HDPE in grams	Weight of PETE in grams	Weight of newspaper in grams	Weight of beam in grams	Maximum load in kilograms	Efficiency

Removing the Beam from the Mold

Three to four days after the beams have been cast, they should be re-moved from the molds. Use extreme caution when dismantling the molds, since the sand mix will have reached only a small proportion of its ultimate strength. Remove all nails and screws before placing the parts in the area designated by your teacher. Return any undamaged screws or other hard-ware to the proper storage area so that they can be used again.

NOTES   ## *THINKING CRITICALLY*

1. Why is it important to design a mold that can be disassembled easily?

_____

_____

2. Name some other examples of materials that must be allowed to cure for a period of time before they are usable.

_____

_____

Specifications and Tolerances

A factory produces many seemingly "identical" items, but these items are not literally identical. They may vary slightly from one to another in size, shape, thickness, or some other particular measurement. Components are manufactured to required specifications. **Specifications** are detailed descriptions that give all the information needed to manufacture the component. They include the type and amount of material to be used and the shape and size of each part. Specifications usually include a range of acceptable values for particular measurements of that manufactured item. This range is called **tolerance.** On notes, the specification is indicated by the desired measurement ± a tolerance of error. For example, a particular manufactured beam might be guaranteed to support a weight of 3000 kg ± 10 kg. Thus, any one beam from this manufacturer might support a load from 2990 to 3010 kg.

1. The constraints for the composite beams that your class made specified a length of 40 cm. Exchange beams with other teams and use a ruler or tape measure marked in millimeters to measure each other's beams. Record and compare the measurements.

2. What is the range of length measurements? What is the mean (average) length? What is the median? (Median is the midpoint. Half of the measurements are smaller; half are larger.) Record the measurements on a bar and whiskers plot such as the one shown in **Fig. 2-13.**

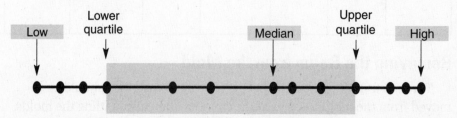

Number line with data points

Fig. 2-12 Sample bar and whisker plot

NOTES

3. How much tolerance do you think should be allowed in the beam construction? Why?

_____

_____

Specifications for one measurement affect other variables in a relationship. To help you understand this concept, refer to the graph of $R = L/B$, **Fig. 2-14.** In this graph, R is the load/beam ratio; L is the load weight; and B is the beam weight, which is constant at 10 kg. The shaded portion of the graph indicates the measurements that fall within the tolerances.

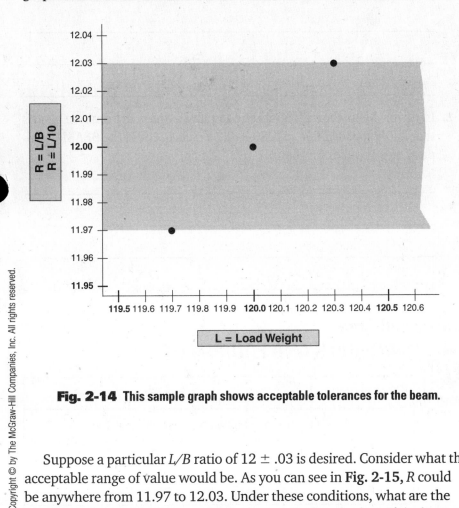

Fig. 2-14 This sample graph shows acceptable tolerances for the beam.

Suppose a particular L/B ratio of 12 ± .03 is desired. Consider what the acceptable range of value would be. As you can see in **Fig. 2-15,** R could be anywhere from 11.97 to 12.03. Under these conditions, what are the possible values of L? What would the range of values be for L if the beam weight is 10 kg? Continue with the fixed L/B ratio of 12 ± .03. What would be the specifications for B if the load weight was fixed at 100 kg?

_____

_____

Project 2 Modeling

NOTES   ***THINKING CRITICALLY***

1. Suppose you work for the BEAM Company. You know that the machine used to mold your beams is capable of producing a beam 6.00 ± 0.50 inches wide. A construction firm has contacted your company and wants to purchase this type of beam for a particular project. This project requires a beam with a width of 6.00 ± 0.35 inches. Can your company provide beams that meet their requirement on a consistent and reliable basis? Explain your answer.

_____

_____

_____

_____

_____

2. In the problem above, if another company wanted to purchase beams that had a specification of 5.80 + 0.70 inches, could the BEAM Company be a supplier for them?

_____

_____

_____

_____

Volume and Area

MATERIALS AND EQUIPMENT

Quantity	Description
1	trough or tank with straight vertical sides
1	metric ruler
as needed	graph paper, 1-cm squares preferred
1	large sheet of butcher's paper

Computing Volume by Displacement

1. You will use this test to calculate the volume of your beam. Work with your teacher to develop a scale for volume that will be placed vertically along the inside of the trough or tank. Keep in mind that the volume of the largest permissible beam is 1,050 cubic centimeters.

2. Ask your teacher to put water in the trough. Record the level of water.

3. Submerge your beam in the trough and record the new level of water. The difference between the two levels is the volume of your beam.

Computing Surface Area of a Cross Section

Your teacher will place all the constructed beams at the front of the room in such a way that everyone can see the beams' cross sections.

1. Use a metric ruler to measure the various beams.

2. Make proportional sketches of the cross sections of the beams. These sketches will be most accurate and easier to do if you use graph paper. You can use graph paper with 1-cm squares, or you can use a simple scale of 1 cm to 1 unit on the graph paper.

3. Determine the surface area of each of the given cross sections.

Determining Volume Analytically

Now that you have computed volume using displacement and also looked at the surface area of the beams' cross sections, you should be ready to find the volume through analytical methods.

1. Suppose the beam was sliced into 1-cm thick pieces parallel with the cross-sectional end, just like a loaf of bread is cut. How could you determine the volume of one of these beam slices?

_____

_____

_____

_____

2. How many of these slices would be in the beam? Measure the beam with a metric ruler to determine this amount. How is the length of the beam (measured in cm) related to the number of slices?

_____

_____

NOTES

Based on the volume of one slice and the number of slices in the beam, how could the volume be computed without using the displacement tank? You could use the formula

Volume = area of a representative cross section × length

This formula is appropriate for solids having uniform cross sections. However, not all solids have this property. A cone is an example of a solid that has similar, but not uniform, cross sections. If you designed a beam in which the cross section at the ends is different from that in the middle, this formula will not work.

3. Calculate the volume of your beam and compare the result to the volume you determined through the displacement of water. Chances are the results will not be the same. What might be the reasons for discrepancies between the two results?

_____

_____

_____

_____

Determining Area of Lateral Surfaces

1. Suppose the outer lateral (side) surfaces of your beam are to be wrapped with a protective cover. Determine the amount of covering needed. Cut that amount from a large roll of butcher's paper. Your goal is to make a covering that will fit exactly with no overlap.

2. Cover your beam with the paper. If there is any overlap, cut it off and tape the excess paper on the chalkboard with your team's name.

3. Each team should compute the area of the excess paper and record this figure on the board.

4. When all teams are finished, exchange beams. Check to see that the beams are covered correctly with no overlap and verify calculations of the area of the excess paper.

Area and Stress

Cross-sectional area influences the stress on your beam. Consider the formula for stress.

Stress = force / area

If force remains constant and area is decreased, will the stress increase or decrease? How will increasing the area affect stress?

THINKING CRITICALLY

1. How could the volume of a right cone be determined with and with-out mathematical calculation? A right cone has its top directly above the center of its base. See **Fig. 2-15.**

_____

_____

_____

_____

2. What happens to the volume of the composite beam if the cross-sectional area is doubled?

_____

_____

_____

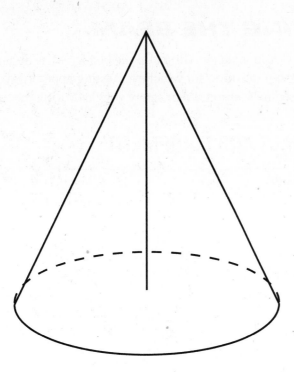

Fig. 2-15 **A right cone**

Project 2 Modeling

NOTES

Evaluation

In the evaluation phase, each team will weigh its beam and then test it to the point of breakage. After the beam has been tested and the necessary data collected, each team will determine the efficiency of its beam. The efficiency is the ratio of how much weight the beam held (the load) to the weight of the beam itself. The composition of the beam will be analyzed as well as its cost.

> ## SAFETY FIRST
>
> **Before You Begin** Make sure you understand how to use the tools and materials safely. Have your teacher demonstrate their proper use. Follow all safety rules. Refer to the Safety Handbook, beginning on page 291, for more information about safety in the lab.

TESTING THE BEAM

Engineering Technology

Your teacher will set up the beam testing apparatus. Follow the procedures for whichever testing apparatus your teacher has set up. Caution: Wear safety goggles during the testing.

MATERIALS AND EQUIPMENT

Quantity	Description
1	scale, 0 to 40 kg or 0 to 10 lbs.
1	beam testing apparatus

Bucket Method

The testing apparatus for the Bucket Method is shown in **Fig. 2-16**. This device uses a five-gallon plastic bucket, such as those used to package ready-mixed drywall compound. The bucket is hung from the beam. Your teacher will gradually fill the bucket with water or sand until the beam breaks.

1. Weigh your composite beam.

2. Enter the weight in the Beam Performance Table.

3. Weigh the sand or water in the bucket when the beam breaks.

4. Record the weight at which the beam broke on the Beam Performance Table.

5. Save the pieces of your beam for later use.

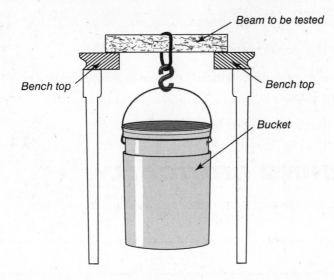

Fig. 2-16 The bucket is filled with sand or water until the beam breaks.

Digital Scale Method

The Digital Scale Method is shown in **Fig. 2-17**. This device uses the readout from a digital bathroom scale and a hydraulic bottle jack. Your teacher will operate the device, causing the jack to press up on the beam until it breaks.

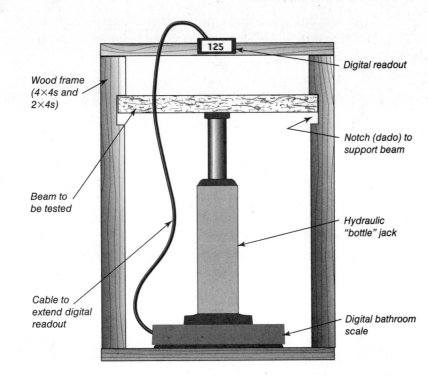

(**Source:** Design by Kent Schweitzer)

Fig. 2-17 A hydraulic jack and a digital bathroom scale are the main components of this beam testing device.

Project 2 Evaluation

NOTES

1. Weigh your composite beam.

2. Enter the weight in the Beam Performance Table.

3. Record the weight at which the beam broke on the Beam Performance Table.

4. Save the pieces of your beam for later use.

THINKING CRITICALLY

1. If you were to redesign your beam, how would you change it?

_____

_____

2. If you had your choice of any materials with which to reinforce your beam while keeping its weight to a minimum, what materials would you choose?

_____

_____

Analyzing Beam Failure

Failure analysis is an important part of engineering design. Knowing how and why a product failed helps engineers design improved versions.

Analyzing Data and Drawing Conclusions

1. Your team should collaborate on a report hypothesizing why your beam failed. At minimum, the report should include the following topics:
   • Why the beam fractured in the place that it did
   • What influence the shape of the beam had on its performance
   • What influence your efforts to reduce the beam's weight had on its performance
   • What influence your efforts to reinforce the beam had on its performance
   • How valid you think the beam testing was
   • What you would do to improve the performance of the beam if you built another one

2. Present an oral summary of your report to the class. Show the pieces of the broken beam.

THINKING CRITICALLY

1. What are some ways in which the results of scientific experiments with structures do not hold true when they are applied to real structures such as bridges?

_____

_____

_____

2. If a beam you designed held 300 kg when it failed, what would you consider when making a decision about how much weight it would hold safely over a long period of time?

_____

_____

_____

_____

_____

Determining Beam Efficiency

Fig. 2-18 shows a sample model using a ratio to compare the weight of the load a beam will hold to the weight of the beam. In engineering terms, this ratio is *the efficiency of the beam.*

Analyzing Ratios

1. Look at Fig. 2-18 and analyze the effects that changes in the numerator and denominator have on a ratio. For example, what if Beam 3 weighs 2 kg and the load at failure was 6 kg? Write that ratio and its decimal equivalent on Fig. 2-18.

2. For Beam 4, suppose the weight is 2 kg and the load is 4 kg. Write that ratio and its decimal equivalent on Fig. 2-18. Do any of the four beams have the same L/B ratio?

	Beam 1		Beam 2		Beam 3		Beam 4	
	ratio	decimal	ratio	decimal	ratio	decimal	ratio	decimal
Weight of Load / Weight of Beam	$\frac{1}{2}$	.5	$\frac{2}{1}$	2				

Fig. 2-18 Sample model for rating success of a beam

N O T E S

3. Look at the sample data in **Fig. 2-19**. Which students' beams have the highest efficiency? Note that efficiency is not determined only by beam weight or only by load. Efficiency depends on the relationship between these two factors.

Student No.	Beam Weight	Load Supported	Efficiency (Load/Weight)
1	2.00	10.0	5.0
2	3.25	11.0	3.4
3	4.25	28.5	6.7
4	2.50	15.0	6.0
5	8.00	32.0	4.0
6	2.00	15.0	7.5
7	3.00	15.0	5.0
8	7.00	7.0	1.0

Fig. 2-19 Sample data

4. In Fig. 2-20, two students' beams weigh the same amount (Students 1 and 6). Why are their L/B ratios different?

_____

5. Suppose you are designing beams in which the desired efficiency is 12. As the beam weight increases/decreases, what does the load weight need to do in order to maintain an L/B ratio of 12? Consider various quantities for L and B so that the L/B ratio remains constant. Then add these values to the graph and rate table in **Fig. 2-20**.

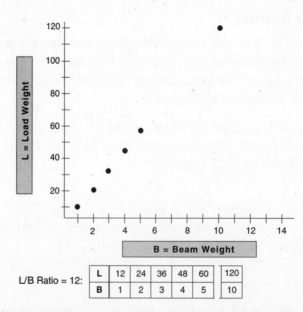

L/B Ratio = 12:	**L**	12	24	36	48	60	120
	B	1	2	3	4	5	10

Fig. 2-20 Graph and weight table for situations in which L/B ratio remains constant

Evaluating the Composite Beams

1. Compute the Load/Beam ratio (efficiency) of your team's beam, using the data you recorded on the Beam Performance Table. Enter ratio in the column Efficiency on the Beam Performance Table.

2. Also enter your data on a large table that is visible to the entire class (or a computer spreadsheet with a projection system). **Figure 2-21** shows a possible table for class data.

Student Team Name	Beam Weight in kg	Load Supported in kg	Load/Beam Ratio	Types of Reinforcement Materials Used	Cross-Sectional Shape of Beam

Fig. 2-21 Suggested table for class entries

3. Why does increasing the load supported not necessarily mean a larger L/B ratio?

_____

4. Why does decreasing the weight of the beam alone not necessarily mean a larger ratio?

_____

_____

5. Why does having a heavier beam not necessarily mean that the beam will support a larger load?

_____

_____

6. Which beam(s) do you think are the "best"? Justify your answer.

_____

7. Analyze your beam. Prepare a written report that addresses how changes could be made to improve the L/B ratio.

8. Using the class data collected in Step 2, prepare the following graphs:
   - Beam weight vs. load weight
   - Beam weight vs. L/B ratio
   - Load weight vs. L/B ratio

NOTES

9. Describe trends you observe in these graphs. For example, do heavier loads correspond to higher L/B ratios (higher efficiency)?

_____

_____

_____

_____

_____

_____

_____

THINKING CRITICALLY

1. The efficiency of the composite beam was determined by L/B ratio. What are some other uses of the term "efficiency"?

_____

_____

_____

_____

_____

2. In what situations might an efficient beam be a disadvantage?

_____

_____

_____

Analyzing Beam Composition

Math   Science

1. With your teacher's help, prepare a large Class Table that combines the data from each team's Beam Performance Table. (Each student team should fill in the data from their Beam Performance Table.) See the example in **Fig. 2-22.** Note that there is an additional column for recording the weight of the concrete.

Project 2 Evaluation

Name_____ Date_____ Class _____

NOTES

Student Team	Weight of Composite Material					Testing Results		
	Aluminum	HDPE	PETE	Newspaper	Concrete	Aluminum	Maximum Load in Kilogram	Efficiency

Fig. 2-22 Sample class table: Your version should be large so that it can be displayed in front of the class.

2. How much do you think the concrete in your beam weighed? You can develop a mathematical model (equation) to answer this question. For example:
   - Total weight of recycled material + weight of concrete = beam weight
   - Therefore, weight of concrete = beam weight - total weight of recycled material

3. Use the information from the Class Table to prepare graphs of each of the following:
   - Weight of high-density polyethylene plastic (for example, milk jugs) vs. beam weight
   - Weight of polyethylene terephthalate plastic (for example, soda bottles) vs. beam weight
   - Weight of newspaper vs. beam weight
   - Weight of metal vs. beam weight
   - Weight of total recycled materials vs. beam weight
   - Weight of concrete vs. beam weight

4. Also prepare graphs for each recycled material vs. load supported.

5. Prepare a written description of each graph. Describe any relationships or trends.

6. Compute the percentage of the total weight of each of the materials (including the concrete) used in the composition of your team's beam.

Project 2 Evaluation

NOTES

7. Recall what you learned about specifications and tolerances (page 000). Imagine that you must construct a beam that has at least 20 percent but no more than 45 percent recycled materials by weight. What is the specification of recycled materials, using desired measurement + tolerance?

_____

8. Suppose you needed to produce a beam with specified percentages of composition. For example, you might make a beam that has specifications of 20% + 2% polyethylene chips and 7% + 1% aluminum by weight, with the remainder in concrete. What are the specifications and tolerances for the concrete in this beam? What would each of these components weigh if the beam weighs 6 kg?

_____

_____

_____

_____

_____

9. Using information from the Class Table, determine the ratio of HDPE plastic chip weight to beam weight and to load weight. Do the same for the other recycled materials.

10. Suppose you are analyzing a beam made of concrete and metal. If the ratio of metal weight to supported weight is 3 to 16, and the beam supported 24 kg, how many pounds of metal are in the beam? If the ratio of metal weight to beam weight is 3 to 8, what percent of the beam is metal?

_____

_____

THINKING CRITICALLY

1. The efficiency of the beam was defined mathematically as the ratio between the load the beam carried to the weight of the beam. What other types of efficiency can be expressed mathematically?

_____

_____

_____

_____

Project 2 Evaluation

NOTES

2. Is it possible to determine what percent of metal is in the beam if the ratio of metal to plastic chips in a beam is 3 to 2? Why or why not?

_____

_____

_____

Curing Time and Strength

Math

1. The graph in **Fig. 2-23** reflects curing of concrete under normal circumstances. What relationship exists between curing time and strength? How many days until the concrete reaches 50% of its ultimate strength? When does it appear that the concrete would be near (at least 90%) its ultimate strength?

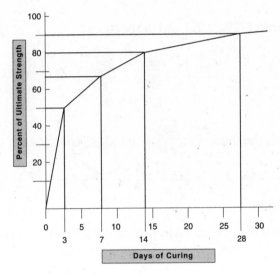

Fig. 2-23 Strength of concrete in relation to curing time

_____

_____

2. You may have seen a crumbling concrete wall or a highway strip that was breaking down. Over time, concrete will begin to break down if exposed to certain factors in the environment. What do you think the life expectancy of your beam would be under normal outdoor conditions? What sources did you use to get this answer? What factors will influence the life span of a beam? Why?

_____

_____

Project 2 Evaluation

3. How many days did your beam cure? According to Fig. 2-23, at what percentage of full strength was your beam on the day of testing?

_____

4. Use the answer from Step 3 to figure out what the beam's full strength would be theoretically. Here are two examples to help you solve this problem. In the examples, the beam was tested on day 7, when it was 65% of ultimate strength, and the beam held 100 lb.

   Example 1: Use the formula:

   percent strength $\times$ potential ultimate strength = tested strength
   Thus,
   $$65\,x = 100$$
   $$x = 153.85$$

   Example 2: Use ratio and proportions:

   $$\frac{.65}{100} = \frac{100}{x}$$
   $$x = 153.85$$

   What is the theoretical full strength of your beam?

   _____

   _____

5. Suppose your beam must support 50 kg and you know that on day 14 an identical beam supported 37 kg. Will your beam meet specifications if allowed to cure a full 28 days?

   _____

   _____

6. Tom was using Fig. 2-23 to analyze the strength of the beam he made. He said that on day 5 the beam would support about 60 lbs. and on day 15 it would support a little over 80 lbs. Since he tested his beam on day 10, he believed his beam should have supported about 70 lbs. Explain his error. How would you help him correct his thinking?

   _____

   _____

Project 2 Evaluation

THINKING CRITICALLY

1. What are some of the reasons why a beam might not reach its predicted ultimate strength during testing?

_____

_____

_____

_____

_____

2. You used sand mix to cast your "concrete beam." Sand mix consists of sand and cement. Actual concrete consists of sand and cement plus gravel. Why is gravel used?

_____

_____

_____

_____

Cost Analysis

Ask your teacher to tell you the weight of the sacks of sand mix (usually 60 lbs. or 90 lbs.) and the number of sacks that were used for making the beams.

Obtain the cost of the sand mix and the costs of the recycled materials from your teacher. If these costs are not readily available, here are approximate costs of the materials in their raw, unprocessed form:

- Sand mix: $4.80 per 90-lb. sack
- High-density polyethylene: $2.15 per pound
- Polyethylene terephthalate: $3.90 per pound
- Aluminum: $2.40 per pound
- Newsprint: $.50 per pound

1. The sand mix that you used in constructing your beam is sold by the sack. Each sack yields a specified volume of mixed sand mix; a 60-lb. sack yields ½ cu. ft. and a 90-lb. sack yields ⅔ cu. ft. Determine the cost per cubic centimeter of the sand mix used to cast the beams.

2. Once you have determined the cost per cubic centimeter, compute the cost of the sand mix used to make your team's beam. Hint: Your beam was made with 5 lbs. of sand mix.

Project 2 Evaluation

NOTES

3. Compute the cost of the other materials in your team's beam. Hint: Use the weight data in the Class Table.

4. Compute what percentage each material contributes to the total cost of the beam.

5. Graphically analyze the relationship between total cost and the volume of each component. As a particular component's volume is increased in a beam (and thus, the volume of another component decreases), what happens to the costs associated with the beam?

6. Compare the *L/B* rating (efficiency) with the total cost of the beam. Graph the class data of beam cost vs. *L/B* rating. Is there any relationship between the *L/B* rating and the cost? Do cheaper beams have higher ratings? Do large amounts of recycled material produce cheaper beams? Is there a trade-off in cost and weight supported?

THINKING CRITICALLY

1. Engineers have to make decisions about the design of structures that rely on mathematics. Mistakes can result in loss of life. What are some of the ways in which the accuracy of the mathematics can be better assured?

_____

_____

_____

_____

2. If you decided to build a concrete bridge twice as strong as you determined it needed to be, would the cost to build it double as well?

_____

_____

_____

_____

Project 2 Evaluation

Cabin Insulation

Design Brief

Scout troops in your area have access to a camp in the woods. The cabins in this camp are fine for summer camping. However, the scoutmasters want to take troops there in the winter as well. See **Fig. 3-1**. Unfortunately, the cabins get very cold at night in the winter. So, the scoutmasters have decided to insulate the cabins. To keep costs down, they will decide to use recycled food-packaging materials for the insulation, since there is a good source for these materials near the camp. Now they must determine which recycled materials might work best for cabin insulation and how panels might be constructed from these recycled materials to maximize their insulating properties.

Fig. 3-1 Insulation can help cabins stay warm in the winter.

Challenge

Design and construct an effective insulation panel using recycled food-packaging materials in the insulation layer. Then test the insulation panel for effectiveness.

Criteria

- You will test your insulation panel for its insulating ability by attaching it to the top of a testing box. The box will contain a light bulb that serves as a heat source. The light bulb will remain on until the temperature in the box reaches 40°C above room temperature, at which point the light bulb will be turned off. A record will be made of the time that it takes for the temperature inside the box to cool to 10°C above room temperature. The insulation panel that delays heat loss for the longest time will be considered the best solution to this problem.

NOTES

- All designs must be accompanied by appropriate documentation, including the following:
  - ☐ A list of all the materials that you considered for your insulation panel
  - ☐ A paragraph describing how you reached the decision to use the material that you selected
  - ☐ Sketches of possible designs
  - ☐ Charts and graphs produced from the testing of the insulation panels
  - ☐ All other related technology, science, and mathematics work completed during the activity
  - ☐ Notes made along the way

Constraints

- Your solution must use only recycled food-packaging materials in the insulation layer.
- Your panel must fit within a wood frame that has inside dimensions of 3 cm × 30 cm × 30 cm (1¼″ × 12″ × 12″).

Engineering Design Process

1. **Define the problem.** Write a statement that describes the problem you are going to solve. The Design Brief and Challenge provide information.
2. **Brainstorm, research, and generate ideas.** The Research & Design section, beginning on page 61, provides background information and activities that will help with your research.
3. **Identify criteria and specify constraints.** Refer to the Criteria and Constraints listed on page 59 and 60. Add any others that your teacher recommends.
4. **Develop and propose designs and choose among alternative solutions.** Remember to save all the proposed designs so that you can turn them in later.
5. **Implement the proposed solution.** Once you have chosen a solution, determine the processes you will use to make your insulation panel. Gather the tools and materials you will need. Make sure you understand and follow all safety rules.
6. **Make a model or prototype.** The Modeling section, beginning on page 74, provides instructions for building the panel.
7. **Evaluate the solution and its consequences.** The Evaluation section, beginning on page 77, describes how to test and analyze the insulation panel.
8. **Refine the design.** If instructed by your teacher, make changes to improve your design.
9. **Create the final design.** If instructed by your teacher, improve the existing panel or make a new panel based on the revised design.
10. **Communicate the processes and results.** Be sure to include all the documentation listed under Criteria.

Research & Design

Have you noticed how, during cold weather, a room gradually grows cooler until the heating system turns on again? This is because heat energy scatters and then escapes to the outside to be displaced by cold air. The purpose of insulation is to limit the loss of heat and prevent cold air from coming in. To design the most effective types of insulation, engineers must have a thorough understanding of temperature measurement and heat. For this project, you too will learn about them. You will conduct experiments on how heat flows through conduction, convection, and radiation and how it is lost. You will then analyze the data collected. After examining examples of materials used in insulation, you will compare their effectiveness by means of the commonly used R-value. You will also learn about various insulating techniques.

HEAT FLOW

Heat energy flows between materials when one material is warmer than the other. The materials may be solids, liquids, or gases. For example, heat will flow between two pieces of metal even if the pieces are not in direct contact. If they are separated, the heat first flows into the air surrounding the warmer piece of metal, and then from the air to the cooler piece. In fact, there are three ways that heat can flow: conduction, convection, and radiation. See **Fig. 3-2**.

Conduction   **Convection**   **Radiation**

Fig. 3-2 The three types of heat flow include conduction, convection, and radiation.

NOTES

Conduction

Conduction refers to the movement of heat through the molecules of a material by means of direct contact. As those molecules in direct contact with heat (such as an open flame, hot water, or a heated surface) become warm, they pass heat energy to neighboring molecules. In turn, these molecules pass heat energy to their neighboring molecules, and so on. Heat passes from molecule to molecule at different rates in different materials. For example, copper conducts heat more readily than plastic does. The following experiment demonstrates how heat is conducted through different materials.

MATERIALS AND EQUIPMENT

Quantity	Description
assorted	rods made of various materials but of equal dimensions, approximately 1 cm in diameter and 20 cm long
1	cake of beeswax
1	wide beaker, less than 20 cm tall

SAFETY FIRST

Before You Begin Do this experiment only with your teacher's permission and supervision. Keep in mind that hot water can cause very serious burns.

1. Apply a small but equal amount of beeswax to one end of each of the rods.

2. Place the rods, beeswax end up, into a beaker of hot water. Be sure that the end with the beeswax is *not* in the water.

3. Observe on which rod the beeswax melts first, second, and so forth. Record your results on the lines below.

_____

_____

_____

_____

_____

_____

Convection

Another way heat energy is transferred is by **convection.** Heat is transferred from warmer to colder areas by the movement of a fluid, such as air or water. Because cool fluids are usually denser than warm fluids, gravity makes the cool fluids sink. This forces the warm fluids upward, displacing them. Movements in fluids caused by heat convection are called "convection currents." Wind and ocean currents are examples of convection currents. The following experiment demonstrates convection.

MATERIALS AND EQUIPMENT

Quantity	Description
1	wide beaker
1	tripod or other stand for beaker
1	heat source (Bunsen burner or candle)
1 bottle	1 food coloring, preferably a dark color such as red or blue

SAFETY FIRST

Before You Begin Do this experiment with your teacher's permission and supervision. Follow safety rules for using a Bunsen burner or a candle.

1. Fill a beaker with cold water and place it on a tripod.

2. Allow the water to stand undisturbed for a few minutes so that its temperature reaches equilibrium.

3. Light a Bunsen burner or a candle and place it under one side of the beaker with cold water.

4. After the water has been heated for about a minute, carefully place one or two drops of food coloring in the water on the side *opposite* the heat source, disturbing the water as little as possible.

5. What happens to the food coloring? Record your observations below.

_____

_____

Radiation

The third way heat is transferred is by **radiation**, in which waves of electromagnetic energy spread out (radiate) through air or empty space. When the waves strike an object, the object absorbs this radiant energy and becomes warmer.

NOTES

Convection and conduction require matter for heat energy to be transferred. Radiation does not. For example, heat from the sun is radiant energy. This heat flows through the vacuum of space before it reaches the earth's atmosphere. (Since space is a vacuum, the heat energy cannot be transferred by convection or conduction.)

Heat radiation is not necessarily visible. For example, many radiant heat sources emit infrared rays, which our eyes cannot normally see. You may have seen satellite pictures of forests or cities in which different colors are used to represent different temperatures. Such pictures are called thermograms. They indicate the relative amount of infrared light emitted by an object.

Which do you think absorbs more heat radiation: dark-colored objects or light-colored objects? Try the following experiment.

MATERIALS AND EQUIPMENT

Quantity	Description
2	containers, one painted black and one painted white
1	thermometer

1. Fill the black and the white containers with equal amounts of water.

2. Record the temperature in each container.

_____

_____

3. Place both of the containers in the sunlight or near a heat lamp for several minutes.

4. Record the temperature of the water in each of the containers again.

_____

_____

THINKING CRITICALLY

1. Why is insulation used in an attic thicker than insulation that is used in the walls of a house?

_____

_____

_____

_____

2. Can heat be transferred through matter by radiation? If so, give an example of this.

NOTES

_____

3. After snow falls on a black asphalt driveway or street, the snow near areas of still-exposed asphalt melts more quickly than do patches of snow on frozen grassy areas. Why does this difference occur?

_____

_____

_____

4. "Sun tea" is made by putting bags of tea in a container of water and placing it in the sun. This eliminates heating the tea on a stove. How might an experiment be designed to determine the best container for making sun tea?

_____

_____

_____

_____

Controlling Heat Flow

Science

When you make your cabin insulation panel, you will want to use materials that resist the flow of heat. Your teacher will distribute samples of various materials. Evaluate those materials using a scale of ten, with ten being the most heat resistant. Which materials do you think will resist the flow of heat best?

Look at the table in **Fig. 3-3** on page 66 showing R-values. **R-value** is a measure of a material's resistance to heat flow (thermal resistance). The greater the R-value, the more resistant the material is. Materials with a high R-value make good insulators. How do the values you assigned to the sample materials compare to the values listed in the table?

In Fig. 3-3, note the R-value of air. Gases, in general, are poor conductors of heat and make good insulators. Compare that to the low thermal resistance of glass. Glass offers very little resistance to heat flow and makes a poor insulator. This is why newer windows have double or triple panes of glass with an inert gas between the panes. The gas helps make the windows more resistant to heat flow.

Small pockets of air are contained in insulation materials. Examples include plastic foam and glass-fiber blanket insulation. (Note that in Fig. 3-3, polyurethane foam has the highest R-value.) Think about ways in which you could incorporate air pockets into the design of your insulation panel.

Name_____ Date_____ Class_____

R-Value of Selected 1″-Thick Materials	
Material	**R-Value**
Air, no convective heat currents, average value	1.00
Concrete and stone	0.08
Glass fiber, insulation batts	3.13
Glass, window, single pane	0.14
Hardboard, standard	1.24
Hardboard, tempered	1.00
Plywood, softwood	1.24
Polystyrene foam, bead board	3.85
Polystyrene foam, extruded	5.00
Polyurethane foam	7.00
Wood, softwood	1.25
Wood, hardwood	0.91

Source: *Handbook of Fundamentals*, American Society of Heating, Refrigerating, and Air Conditioning Engineers

Fig. 3-3 Thermal resistance (R-value) of selected 1-inch thick materials

Refer to **Fig. 3-4** showing a vacuum bottle. Large vacuum bottles (often referred to by the brand name Thermos®) are constructed of two containers, one inside the other. The air between the containers is removed, creating a vacuum, and the space sealed. The vacuum prevents heat flow by conduction, since there is no material through which the heat can flow. It also prevents convection, since convection currents require a fluid. The walls of the inner container are covered with a reflective material, which reduces radiation by reflecting the heat away from the walls and back into the coffee or other stored substance.

Many houses are subject to the "chimney effect." This occurs when heated air flows out the top of chimneys or through gaps around upper windows or other parts of the house. Cold air then flows in through gaps in the lower part of the house to displace the warm air that has been lost. Creating tight seals around the house by using storm windows and weather stripping can reduce the chimney effect. For similar reasons, make sure that your cabin insulation panel fits snugly against the top of the testing box so that the heated air inside does not escape.

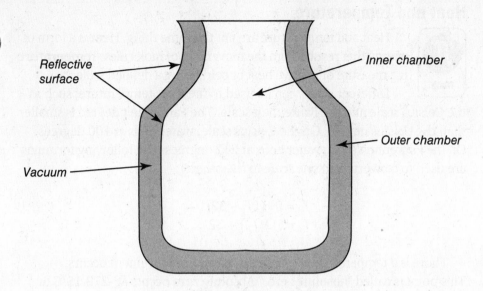

Fig. 3-4 A Dewar flask, a common type of vacuum bottle

THINKING CRITICALLY

1. What are some ways in which homes might be made to hold heat more efficiently?

   _____
   _____
   _____
   _____
   _____

2. How good an insulator is the soil that surrounds a house's foundation?

   _____
   _____
   _____
   _____
   _____

3. Why do basements remain at a relatively constant temperature all year long?

   _____
   _____

NOTES

Heat and Temperature

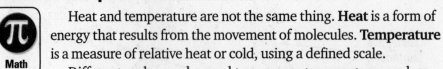

Heat and temperature are not the same thing. **Heat** is a form of energy that results from the movement of molecules. **Temperature** is a measure of relative heat or cold, using a defined scale.

Different scales can be used to measure temperature, such as the Celsius scale and the Fahrenheit scale. The Fahrenheit degree is smaller than the Celsius degree. On the Celsius scale, water boils at 100 degrees. On the Fahrenheit scale, water boils at 212 degrees. The following formulas are used to convert from one scale to the other:

$$T_C = 0.6\,(T_F - 32)$$
$$T_F = (1.8\,T_C) + 32$$

There is a temperature at which no molecular movement occurs. This point is called "absolute zero." Absolute zero occurs at -273.15°C or -459.67°F.

The kelvin scale is also used to measure temperature. A kelvin is the same size as a Celsius degree, but the scale begins at absolute zero. On the kelvin temperature scale, absolute zero is 0 K, and water boils at 373.15 K.

Heat energy is not measured in temperature units. Today it is usually measured in joules (J), which is the amount of work done by one newton acting over a distance of one meter. However, other units, such as the calorie, are also associated with heat. In the past, heat was measured relative to its ability to raise the temperature of water. One small calorie (cal), or gram calorie, is the amount of heat needed to raise one gram of water by 1°C. In the British system, the British thermal unit (Btu) was used. One Btu is the amount of heat needed to raise the temperature of one pound of water 1°F. The relations among the various units are as follows:

$$1\,J = 0.239\ cal$$
$$1\ cal = 4.18\ J$$
$$1\ Btu = 1055\ J = 252\ cal$$

THINKING CRITICALLY

1. Can heat be added to a substance without causing its temperature to rise? If so, give an example. If not, explain why not.

_____

_____

2. Why do you suppose water has been used in the past as the medium for measuring heat?

_____

_____

Temperature Change

Math

For the following experiment, your teacher will divide the class into teams. Each team will test the insulating value of various thicknesses of newspaper on one can of hot water.

MATERIALS AND EQUIPMENT

Quantity	Description
1 per team, plus 1 as a control	metal can, such as a soup can
enough to fill all cans	hot water (cooler than 55°C)
1	ruler
1	bulb thermometer
1 stack	discarded newspapers
1 roll	transparent tape
1	stopwatch
1 sheet per student	graph paper

SAFETY FIRST

Before You Begin There is no need to boil the water for this experiment. Very hot tap water should work well enough. Keep in mind that hot water can cause serious burns.

1. Apply layers of newspaper around the outside of the metal can, using transparent tape to hold the paper in place. Each team should apply a different thickness, such as ⅛″, ¼″, ⅜″, and so on. (You might also count the number of sheets used.)

2. Predict the results of the experiment. Record your predictions.

_____

_____

_____

_____

3. Design a graph for recording the elapsed time and temperature data that will be collected during the experiment. See **Fig. 3-5** on page 70 for an example. Determine how often measurements should be taken (for example, each minute, every two minutes, etc.).

Name_____ Date_____ Class _____

NOTES

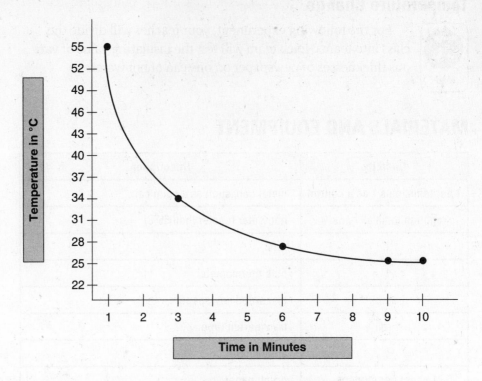

Fig. 3-5 A sample graph

4. Note the room temperature (as indicated on your thermometer) before the can is filled.

5. Fill the can with hot water.

6. Immediately place the thermometer in the filled can.

7. Periodically, monitor the water temperature until the water reaches room temperature. Each time, note the time and temperature data in the table you designed.

8. Each team member should prepare a graph of the data.

9. Analyze your data/graph, draw conclusions, and communicate these conclusions in a brief paragraph.

_____

_____

_____

_____

_____

_____

10. Compare your written conclusions with those of your teammates and with the conclusions of students on other teams. Also, compare your conclusions to the predictions you made earlier.

NOTES

_____

_____

_____

_____

_____

THINKING CRITICALLY

1. Was most of the heat lost from the water early in the experiment or toward the end?

_____

_____

_____

_____

2. Was the time interval you set for collecting the data an appropriate time interval? Why or why not?

_____

_____

_____

_____

3. What should be done about datum points on the graph that do not seem to fall into line (e.g., "outliers")?

_____

_____

_____

_____

_____

NOTES

DESIGNING THE INSULATION PANEL

As you think about your panel design, read the Design Brief, Challenge, Criteria, and Constraints. Then read the list of materials that you will have available to make your panel. Note that dimensions are specified so that your frame will fit within the testing box. You will need to select appropriate materials and make decisions about the way they will be arranged within the frame. One possible design is shown in **Fig. 3-6**.

Cross Section of Insulation Panel

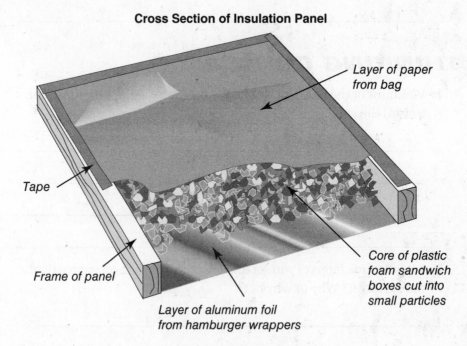

Layer of paper from bag

Tape

Frame of panel

Layer of aluminum foil from hamburger wrappers

Core of plastic foam sandwich boxes cut into small particles

Fig. 3-6 Sample solution

Have you ever seen insulation material for sale in a building supply store? It comes in four main varieties: rigid boards, loose fill, spray-foam, and flexible blankets.

Rigid board insulation is just what it sounds like—a stiff, board-like panel of insulating material. Loose fill is lightweight and resembles shredded newspapers when blown into place. Spray-foam insulation foams up like shaving cream and is sprayed inside walls by using special equipment. Flexible insulation is typically made of plastic-wrapped fiberglass and resembles a thick, narrow blanket.

Which type do you think the insulation in your panel should most resemble? You may want to research that type on the Internet to learn a little more about how it is installed. Some of the information you find may be helpful when making your own design.

As you work, consider what you have learned about the principles of heat flow. Visualize how heat will flow within the testing box shown in Fig. 3-9. Consider the following questions:

NOTES

- Can the fact that darker colored surfaces absorb more radiant energy than lighter colored surfaces be applied to improving the design of the insulation panel?
- Is there a way in which convection currents can be minimized so that the heat energy is less likely to flow through the panel?
- What materials are less likely to conduct heat energy through the panel?

Prepare sketches of several panel designs. Include ideas about how the insulating material should be prepared. Evaluate the sketches and choose the one that you think will work best.

THINKING CRITICALLY

1. If most insulation makes use of trapped air, where is the air located in rigid panel insulation?

_____

_____

_____

2. How could accuracy in dimensioning the panel frame influence the results of the experiment?

_____

_____

_____

3. What is the purpose of weather stripping?

_____

_____

_____

NOTES

Modeling

This section provides instructions for constructing the insulation panel. After selecting the best design for your panel, you and your teammates will construct it. You will begin by designing and building a frame. The recycled materials you have decided to use for insulation will then be placed in the frame.

SAFETY FIRST

Before You Begin Make sure you understand how to use the tools and materials safely. Ask your teacher to demonstrate their proper use. Follow all safety rules. Refer to the Safety Handbook, beginning on page 291, for more information about safety in the lab.

CONSTRUCTING THE INSULATION PANEL

Keep in mind that the frame must be true so that it makes a good seal with the insulation test box. The joints, too, must be tight so that no heat is lost through them.

MATERIALS AND EQUIPMENT

Quantity	Description
54 inches	$\frac{3}{4}'' \times 3$ cm pine stock or plywood
8	4d box nails
1	hammer
1	handsaw
1	stapler or roll of tape
2 sheets	33 cm $\times$ 33 cm Kraft paper
1	straightedge
1	table saw (optional)

1. With your teacher's permission, use a handsaw or table saw to cut the boards to the proper size. Refer to **Fig. 3-7**.

2. Construct the frame for your panel as specified in your design sketch.

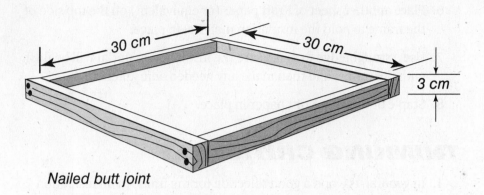

Nailed butt joint

Fig. 3-7 Details of the insulation panel frame to be constructed from
¾" pine stock or plywood

3. Staple or tape Kraft paper (or the equivalent) to the bottom of
   the frame.

4. Before you place the insulating materials in the frame, look at
   Fig. 3-8. The top and bottom of the insulation panel must be flush
   with the top and bottom of the frame. You can check this with a
   straightedge placed across the edges of the frame so that it touches
   both sides.

5. Install the insulation material.

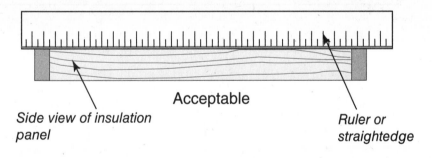

Acceptable

*Side view of insulation
panel*

*Ruler or
straightedge*

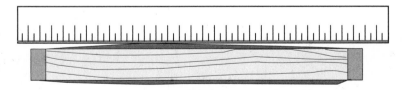

Unacceptable

Fig. 3-8 Testing the insulation panels to make sure they are
the same thickness as the frame

Project 3 Modeling

NOTES

6. Place another sheet of Kraft paper (or equivalent) on the top side of the frame to hold the insulation material in place.

7. Use a straightedge to make sure the insulation material will be flush with the frame, and then make any needed adjustments.

8. Staple or tape the Kraft paper in place.

THINKING CRITICALLY

1. In general, is wood a good selection for the frame of an insulation panel? Why or why not?

_____

_____

_____

_____

2. Why is it important that the top of the insulation be flush with the top of the frame?

_____

_____

_____

_____

Project 3 Modeling

Evaluation

During evaluation, you will place your insulation panel in the testing box shown in **Fig. 3-9**. Heat from the light bulb inside the box will radiate outward to the box walls and to your panel. Conduction will occur when molecules, that make up the walls of the box and the frame of your panel, transfer heat. Convection currents will distribute the heat evenly. They will also cause warm air to flow to the outside and allow cooler air from outside to flow in. Because the box and panel are not air tight, some heat loss will occur. However, any gaps between the box and your panel will speed up this heat loss.

SETTING UP THE TEST

Engineering Technology

Temperature probes inside the box will generate data. During the testing process, you will monitor the time and temperature. Each test takes about 20 minutes. You will record test data during this time on the Insulation Panel Performance Table. After the test, you will prepare a graph, analyze the data, and make conclusions.

SAFETY FIRST

Before You Begin Make sure you understand how to use the tools and materials safely. Ask your teacher to demonstrate their proper use. Follow all safety rules. Refer to the Safety Handbook, beginning on page 291, for more information about safety in the lab.

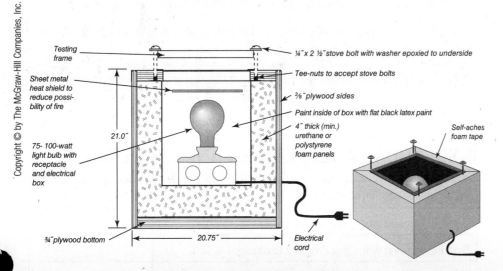

Testing frame

Sheet metal heat shield to reduce possibility of fire

75-100-watt light bulb with receptacle and electrical box

21.0˝

¾˝ plywood bottom

20.75˝

Electrical cord

¼˝ x 2 ½˝ stove bolt with washer epoxied to underside

Tee-nuts to accept stove bolts

⅜˝ plywood sides

Paint inside of box with flat black latex paint

4˝ thick (min.) urethane or polystyrene foam panels

Self-aches foam tape

Fig. 3-9 Details of insulation testing box   Pictorial View

Project 3 Evaluation

Name_____ Date_____ Class_____

NOTES

MATERIALS AND EQUIPMENT

Quantity	Description
1	insulation panel testing box
2	temperature probes with computer interface and software

1. Your instructor may ask you to help build the testing box. If so, refer to Fig. 3-9 on page 77 and follow his or her instructions. If not, examine the box to be used. Observe the heavy insulation on the walls, the light bulb that will act as a heat source, and the spot where your insulation panel will be placed. Although the box is heavily insulated, some heat will be lost.

2. Study the flow chart shown in **Fig. 3-10**. The chart outlines the steps you will follow during the testing process.

3. Each panel must be tested under similar conditions. Before the first test, and if the box has grown cool between tests, the box must be preheated. To do this, cover and heat the testing box up to the testing temperature (40°C above room temperature). Then, turn off the heat source and wait 5 minutes before inserting a panel.

4. Install the insulation panel in the preheated box, and turn on the light bulb to raise the temperature inside the box back to 40°C above ambient room temperature. Then turn off the light bulb.

5. Monitor the time that it takes for the temperature inside the box to drop to 10°C above ambient room temperature. Note the time and temperature data in 3-minute intervals. **Fig. 3-11** on page 80 shows an example of how this can be done. If you are using a temperature probe system that interfaces with a computer, you will not have to monitor the temperature yourself. After the test, you will be able to print out a table showing time and temperatures using the program's software. In either case, save your data for later use.

6. When the test is finished, record the data in the Insulation Panel Performance Data Table below.

7. The panel that holds the heat above the designated temperature for the longest time is the *best* insulator.

INSULATION PANEL PERFORMANCE DATA TABLE

Initial ambient (room) temperature	
Time (in minutes) when box reached ambient + 40°C	
Time (in minutes) when box dropped to ambient + 10°C	
Difference between t_1 and t_2 in minutes	

Project 3 Evaluation

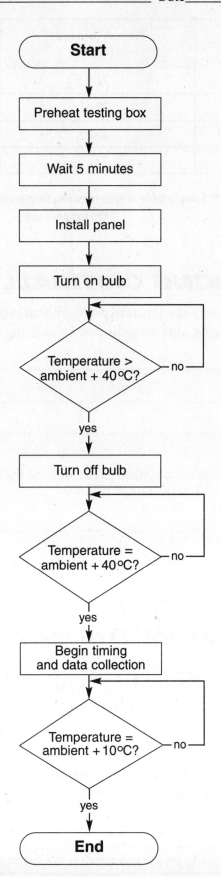

Fig. 3-10 Flow chart of testing procedure for insulation panel

Project 3 Evaluation

NOTES

Elapsed Time	Probe A (ambient) in Degrees	Probe B (inside box) in Degrees
00:06:08	22.1	48.9
00:06:10	22.3	48.9
00:06:12	22.2	47.8
00:06:14	22.2	47.2

Fig. 3-11 Sample table of data resulting from monitoring the elapsed time and the temperature

THINKING CRITICALLY

1. What are some practical problems that might occur if discarded food packaging were actually used for building insulation?

_____

_____

_____

2. Caulk has been applied to the joints of the testing box. Where is it used in a typical house? Why?

_____

_____

_____

GRAPHING THE DATA

Graphs can be very useful when trying to compare test data. In this section you will use your test data and Insulation Panel Performance Table to make a graph. Then you will draw conclusions based on the data.

MATERIALS AND EQUIPMENT

Quantity	Description
1	Insulation Panel Performance Table
1	copy of your test data, like that shown in Fig. 3-11
as needed	graph paper

Project 3 Evaluation

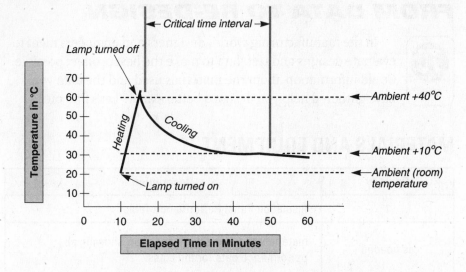

Fig. 3-12 Sample graph of temperature vs. elapsed time

1. Use your Insulation Panel Performance Table to prepare a graph of the data you collected during the testing of your insulation panel. See **Fig. 3-12** for an example.

2. Based on your data and graphs, what conclusions have you drawn regarding the design of insulation and the transfer of heat?

_____

_____

THINKING CRITICALLY

1. What event occurred when the slope of the line on your graph was positive?

_____

_____

2. What event occurred when the temperature stopped increasing and began to decrease?

_____

_____

3. Why was the ambient (room) temperature represented by a horizontal line on your graph?

_____

_____

Project 3 Evaluation

NOTES

FROM DATA TO RE-DESIGN

Math

In the manufacturing world, designers and engineers meet to evaluate designs and test data to make the best product possible. Could information about the materials used and the data you have gathered help you design a better panel? Let's find out.

MATERIALS AND EQUIPMENT

Quantity	Description
1	Insulation Panel Performance Table
as needed	materials to make a large bar chart showing all performance data for the class

1. As a class, develop a bar chart that gives an overview of the insulation panel performance for a teams. The entire class should be able to view the chart at one time. You might want to use an overhead projector, wall chart, or chalkboard. Use the data on the Insulation Panel Performance Tables that you prepared when testing your panels. The chart should include the elapsed time for heat loss, the materials used by each time, and any other features in each design.

2. Based on the data shown in the bar graph, what conclusions can you draw regarding the design of insulation panels?

THINKING CRITICALLY

1. Which material was the best insulator?

_____

_____

_____

_____

_____

2. Which panel held heat the longest? Why? What properties or design features did it have that could be responsible?

_____

_____

_____

_____

Wind-Powered Generator

Design Brief

Wind has been a valuable source of energy for many centuries. Sailing ships depended on the wind to propel them across the oceans. Windmills, large wheels turned by the wind, were originally used to pump water from wells. Modern windmills, grouped into large wind "farms," are now being used by some communities as an alternative source of energy. See **Fig. 4-1**. As the windmills turn, wind energy is converted into electrical energy. Because wind is free and readily available, engineers continue to develop more efficient ways to collect and convert wind energy.

Fig. 4-1 Windmills are being used as backup sources of electricity in some cities. How might geography affect the location of windmills?

Challenge

Design and build a device that collects wind energy and transforms it into the largest possible amount of electrical energy.

Criteria

- Your device must use wind to power a generator and produce electricity.
- Your device must be able to "seek the wind." In other words, it must be able to capture wind coming from any direction.
- During the test, your generator will be placed 15 cm (6″) away from a fan that will produce wind. The fan will be moved to more than one location, the electrical energy generated by your device measured, and an average taken. The device that generates the highest average energy reading will be the winner of the challenge.
- All designs for devices must be accompanied by appropriate documentation, including:

NOTES

☐ Sketches of all the possible solutions you considered
☐ A finalized drawing of your chosen solution
☐ Data collected from various tests
☐ A chart/graph showing how your solution performed
☐ Information gathered from resources
☐ Notes made along the way

Constraints

- Your device must be no more than 2,000 cubic centimeters (122 cubic inches) in volume.

Engineering Design Process

1. **Define the problem.** Write a statement that describes the problem you are going to solve. The Design Brief and Challenge provide information.
2. **Brainstorm, research, and generate ideas.** The Research & Design section, beginning on page 87, provides background information and activities that will help with your research.
3. **Identify criteria and specify constraints.** Refer to the Criteria and Constraints listed on pages 85 and 86. Add any others that your teacher recommends.
4. **Develop and propose designs and choose among alternative solutions.** Remember to save all the proposed designs so that you can turn them in later.
5. **Implement the proposed solution.** Once you have chosen a solution, determine the processes you will use to make your wind-powered generator. Gather the tools and materials you will need. Make sure you understand and follow all safety rules.
6. **Make a model or prototype.** The Modeling section, beginning on page 107, provides instructions for building the wind-powered generator.
7. **Evaluate the solution and its consequences.** The Evaluation section, beginning on page 112, describes how to test the wind-powered generator.
8. **Refine the design.** If instructed by your teacher, make changes to improve your design.
9. **Create the final design.** If instructed by your teacher, make a new generator based on the revised design.
10. **Communicate the processes and results.** Be sure to include all the documentation listed under Criteria.

Research & Design

In this section you will learn about how the force of a wind stream interacts with a surface and how this force can be converted into rotational motion. You will also learn about simple electrical circuits, how electricity is generated using the principles of electromagnetic induction, and how to calculate the volume of a cylinder.

EXPLORING BLADE DESIGN

A gas moving relative to an object exerts a force on that object. Can you think of examples? When the gas is air moving relative to the surface of the earth, we refer to the force exerted on the object as *wind force*. This is the force that holds up a kite or an airplane in flight. It is also the force that makes a windmill go around and makes it possible to get useful energy from the wind.

MATERIALS AND EQUIPMENT

Quantity	Description
1	30 cm square base of ½″ plywood
1	15 cm square slider of ¼″ plywood
assorted pieces	⁵⁄₁₆″ dowel rod cut in various lengths
1	drill
1 sheet	plastic or paperboard
assorted	rubber bands
as needed	glue, sandpaper
1	window fan
1 sheet per team	graph paper
1	protractor

1. Why do you think some windmills might work more effectively than others do?

_____

_____

_____

_____

NOTES

2. Why do you think some windmills have more vanes, or blades, than others might have?

_____

_____

_____

_____

_____

Think about when you have walked into the wind. You can feel the force of the wind pressing against your body. However, if you hold your coat open into the wind with extended arms, you can feel how much the force of the wind against your body increases.

A windmill is really a **turbine**, a machine in which the moving air turns the **blades**, or vanes, to produce rotary motion (mechanical energy). The rotary motion transfers to a generator, where it converts into electricity.

There are two basic types of wind turbines: drag and lift. The drag type is the most familiar. It looks like a multi-bladed fan, and as the wind pulls against the blades, a **drag** force is created. The lift type of windmill operates on the aerodynamic principles of an airplane wing (airfoil). Because airfoils are difficult to create, this project will focus on a drag-type wind turbine.

For this experiment, you will be investigating the wind force on a vane that represents a blade in a wind turbine. The vane could also represent, in a simplified way, a blade in an airplane propeller or the wing of the airplane. You will use the apparatus shown in **Fig. 4-2**. Your teacher may provide a finished apparatus or give you parts for you to assemble.

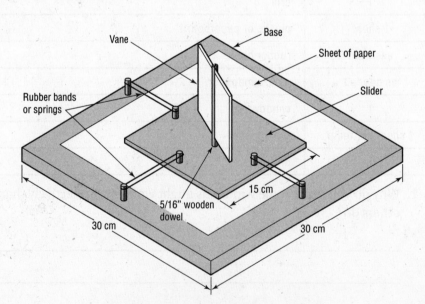

Fig. 4-2 Wind vane experimental apparatus

1. If you and your teammates will be assembling the apparatus, sand the bottom of the slider so that it is smooth to reduce friction.

2. Insert the vane into the slot in the longest dowel and glue it in place. Then push the other end of the dowel into the slider. The fit should be tight.

3. Place the remaining shorter pieces of dowel on the base and slider, as indicated in Fig. 4-2.

4. Tape the sheet of graph paper to the base beneath the slider so that you can mark various positions of the slider.

5. Install the rubber bands. As you work the experiment, try a variety of rubber bands to determine the best selection relative to the movement of the slider in the wind stream.

6. Place the fan 15 cm from the testing apparatus. See **Fig. 4-3**.

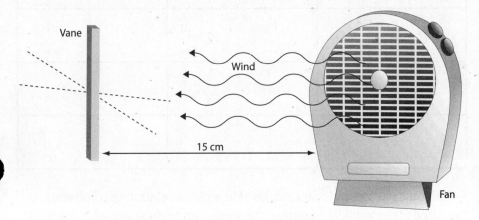

Fig. 4-3 Set-up for wind vane experimental apparatus

7. Set up the testing apparatus so that the vane will be perpendicular to the wind stream. Do *not* turn the fan on yet. Standing behind the apparatus and looking toward the fan, mark on the paper the lower righthand corner of the slider.

8. Using a protractor, measure off 15° increments as you rotate the vane/slider to the left. Mark each incremental position of the slider's lower righthand corner on the paper until the vane is parallel to the wind stream. Next to each mark, write the angle in degrees.

9. Repeat Step 8, rotating the vane/slider to the right.

10. Return the vane to its perpendicular position and turn on the fan.

11. Measure in millimeters from the point of origin the displacement of the vane/slider. Note the displacement on the Wind Vane Displacement Table on page 90.

12. Rotate the vane/slider through the sequence of incremental positions, noting the displacement each time on the table.

Name_____ Date_____ Class _____

WIND VANE DISPLACEMENT TABLE

Angle of Vane Relative to Wind Stream	X Displacement from Origin in *mm*	Y Displacement from Origin in *mm*
90°		
75°		
60°		
45°		
30°		
15°		
0°		

As you have probably noticed, there is a direct relationship between the force of the wind, the angle of the vane, and the distance that the vane moved along the x and y axes.

THINKING CRITICALLY

1. What might be the advantages and disadvantages of wind-generated electricity?

_____

_____

_____

2. Wind reaches higher velocities at higher elevations. What are some of the problems in locating wind generators on top of tall mountains?

_____

_____

_____

BLADE PITCH AND ROTATIONAL SPEED

The total force from the wind acting on a turbine or windmill is the sum of the forces acting on each blade. Several factors should be considered when trying to maximize this force, such as the pitch (angle) of the blades, the size of the area exposed to the wind, and the number of blades. During the following activity, note the relationship between blade pitch and rotational speed.

MATERIALS AND EQUIPMENT

Quantity	Description
assorted	Tinker Toys or other construction toys
assorted pieces	rigid cardboard
1	two-speed fan
1	stopwatch or watch with second hand
1	protractor
1	piece of colored tape or felt tip marker that contrasts with color of cardboard

1. Using the Tinker Toys, construct a simple fan-type wind turbine. See **Fig. 4-4**. Use a pulley with eight holes drilled in its circumference as the hub. The dowels in the set can be inserted into the holes in a variety of positions. The dowels have slots cut in the ends into which cardboard blades can be inserted. The dowels can be turned in the holes so that the pitch of the blades can be varied.

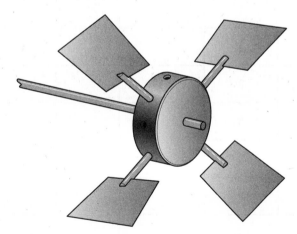

Fig. 4-4 Sketch of windmill turbine made from Tinker Toys

2. Experiment first with four blades. Mark one of the blades with colored tape or some other marker.

Name_____ Date_____ Class _____

3. During the experiment, vary the angle of the blades in increments of 15°. Start with an angle of 0° (parallel to the wind stream) and end with an angle of 90° (perpendicular to the wind stream). A protractor or a worksheet with the required angles drawn on it can be used to set the angles.

4. Position the fan. During the experiment, keep the fan at a constant distance from the turbine and maintain the same fan speed.

5. Measure the turbine's rotational speed (rpm) for the different blade angles this way: Have one team member watch the turbine while another uses the stopwatch to count out ten seconds. The student watching the turbine should count how many times the marked blade revolves around the center during the ten-second period.

6. Record your data in the Rotational Speed Table.

7. Repeat the experiment using eight blades.

ROTATIONAL SPEED TABLE

Pitch Angle	Number of Blades	Rotations in 10 Seconds	RPM	Notes
0°	4			
15°	4			
30°	4			
45°	4			
60°	4			
75°	4			
90°	4			
0°	8			
15°	8			
30°	8			
45°	8			
60°	8			
75°	8			
90°	8			

GRAPHING THE DATA

It is often helpful to look at data you have collected in chart form or some other graphic representation, such as those shown in **Fig. 4-5.** Decide which type of graph would be appropriate to display the data in the Wind Turbine Rotational Speed Table. Display the pitch versus the rpm. Which pitch angle produced the most energy?

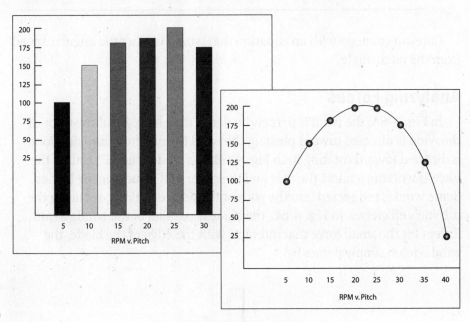

Fig. 4-5 Types of graphs

1. What difference, if any, did changing the pitch make?

   _____

   _____

2. Were you able to predict any differences?

   _____

   _____

3. How might this information help you in designing your wind-capturing device?

   _____

   _____

   _____

NOTES

4. Analyze the data in the chart in order to determine the relationship between pitch, the number of blades, and rotational speed. What conclusions have you reached about that relationship?

_____

_____

_____

Can you come up with an equation that would predict the angular speed from the pitch angle?

Analyzing Forces

In **Fig. 4-6A**, the wind is perpendicular to the blade. All of the force of the wind is directed toward pushing the wind turbine from its axle. None is directed toward rotating it. In **Fig. 4-6B**, the blade is at 45°. Half of the force is working against the axle and the other half is turning the blades. Some wind is redirected into the path of the adjacent blade, reducing the turbine's efficiency. In **Fig. 4-6C**, the blade is parallel to the wind stream. Except for the small force that interacts with the edge of the blade, the wind stream simply passes by.

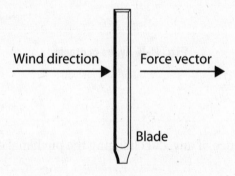

Fig. 4-6A Blade perpendicular to wind

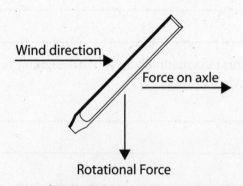

Fig. 4-6B Blade at 45 degrees to wind

Fig. 4-6C Blade parallel to the wind

Which wind turbine configuration produces the optimal performance in terms of speed?

_____

_____

_____

1. Using the optimal set up, experiment with the size and shape of the blades.

2. Move the fan around to simulate changing wind direction. Were you able to predict what happened? How confident were you in the accuracy of your predictions?

Varying Wind Speed

Using the most efficient blade design, hypothesize what will happen if the distance between the fan and the turbine is doubled. What will happen if the fan is operated at half the speed or twice the speed?

1. Set up the turbine as before, but double its distance from the fan. Record the data in the Wind Turbine Variable Wind Speed Data Table.

2. Set up the turbine and fan as for the original test, but increase (or decrease) the fan speed. Record the data in the Variable Wind Speed Data Table.

VARIABLE WIND SPEED DATA TABLE

Distance to Fan	Fan Speed	Rotations in 10 Seconds	RPM	Notes
8 cm	Low			
8 cm	High			
16 cm	Low			
16 cm	High			

Project 4 Research & Design

NOTES

3. What conclusions can you draw about the effect of doubling the distance and halving the speed of the fan?

_____

_____

_____

_____

THINKING CRITICALLY

1. Can you think of other phenomena which, when graphed, would resemble the graph that you created showing pitch angle and rotational speed?

_____

_____

_____

_____

_____

_____

2. What forces inhibit the rotation of the turbine blades?

_____

_____

_____

_____

_____

3. It is theoretically impossible to extract more than 59.3% of the energy from a wind stream. Why do you think this is true? What factors might cause reduced efficiency in the wind turbine?

NOTES

_____

_____

_____

_____

If you try to exceed the 59.3% limit, then air "piles up" in front of the turbine and prevents additional wind from reaching the blades or vanes of the turbine. This figure is ideal and theoretical. In actual practice, a wind turbine is only about 30% efficient.

4. What would happen if your wind turbine were 100% efficient in capturing the wind? Would there be any disadvantages?

_____

_____

_____

_____

_____

_____

5. Give examples of other technologies that must rely on science to deal with the problems associated with wind flow patterns.

_____

_____

_____

_____

NOTES

6. Although the size of the blades potentially provides for better use of the wind's energy, larger blades present other problems. What are some of these problems?

_____

_____

_____

_____

_____

7. If you could vary the pitch of a wind turbine blade while the turbine was running, what would be some of the advantages?

_____

_____

_____

_____

_____

ELECTRICAL INDUCTANCE AND GENERATORS

Science

For this activity you will learn more about how generators work by assembling two circuits. One will contain a battery and a lamp. The other will be based on an electromagnet.

MATERIALS AND EQUIPMENT

Quantity	Description
1	lengths of insulated 22-26 gauge wire
1	Battery
1	lamp to match voltage of battery
1	magnetic compass
1	bar of steel or iron for an induction coil
1	permanent bar magnet
1	Galvanometer
1	open-frame laboratory generator

1. Using the battery, lamp, and wire, assemble a simple electrical circuit.

2. Hold the bar magnet near the filament of the light bulb while the lamp is on. The filament will vibrate. What is happening? The magnet has *induced* the flow of electrons through the wire. In what ways could the amount of current produced be increased?

   _____

   _____

3. Place a magnetic compass near one of the wires of the circuit. Then notice how the current flowing in the wire attracts the needle of the compass.

4. Reverse the polarity of the circuit by switching the leads to the battery. Place the compass near one of the wires again. Notice how the needle moves in the opposite direction.

   A *changing* magnetic field will produce a current in a conductor. The magnetic field *must* be changing in some way or the current will not be produced. This change can be in either intensity (magnitude) or in direction.

5. Wrap wire around the bar of steel to make an electromagnet. Then create a circuit that includes the galvanometer like the one shown in **Fig. 4-7.**

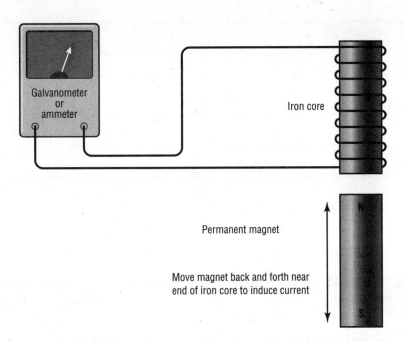

Fig. 4-7 The needle of a galvanometer deflects when a magnet is moved near a coil, showing electromagnetic induction.

Name_____ Date_____ Class _____

NOTES

6. Move the bar magnet around the end of the iron core in various ways in order to get the maximum deflection of the galvanometer needle. What do you conclude from this regarding the orientation of the magnet relative to the amount of current produced and the direction in which it flows?

_____

_____

_____

Your instructor will show you a generator that illustrates what you have learned in this experiment. However, in the generator the coil is moved in relation to the magnet instead of the magnet being moved in relation to the coil. Falling water, steam pressure, or a wind turbine like the one you will build can be used to turn the coil in the generator. The rotating coil is influenced by the changing magnetic field and a current is induced in the coil.

THINKING CRITICALLY

1. Why did the compass needle change directions when the electric current was reversed?

_____

_____

_____

_____

2. Why does a battery connected to a generator rotate?

_____

_____

_____

_____

VOLUME OF A CYLINDER

Math

The wind turbine you will design for this project will have a cylindrical configuration and must be no more than 2000 cubic centimeters (122 cubic inches) in volume. For this activity you will prepare a graph that will help you design your wind turbine so that it meets the required design specifications.

MATERIALS AND EQUIPMENT

Quantity	Description
1 sheet	graph paper
assorted	cylindrical objects
1	computer with spreadsheet software

Keep in mind that you will be working with two variables: the diameter of your wind turbine and its depth. If you change one variable, then the other must change as well to maintain the prescribed volume.

1. Your teacher will provide you with assorted cylindrical objects, such as cans and pipes. Calculate the volume of several of them. The volume of a cylinder is calculated using the formula $V = \pi r^2 h$. As you work, enter your data in the Volume of Cylindrical Objects Table.

VOLUME OF CYLINDRICAL OBJECTS TABLE

Object	Diameter	Height	Circumference	Circular Area	Volume

2. In the formula $V = \pi r^2 h$, r represents the radius of the wind turbine, and the depth of the collector (blades and hub) is represented by h. By assigning given values to h, the diameter of the wind turbine with a constant volume of 2000 cm^3 can be calculated. The expression would be:

$$d = 2\sqrt{\frac{2000}{h\pi}}$$

NOTES

Your teacher will assign your team a cylinder height (or depth). Calculate the corresponding diameter of a wind turbine having a constant volume of 2000 cm^3. Work individually first. Then check your work against that of your teammate.

3. As a class, after the calculations have been made, prepare a table of all the values so that all students can see them and write them down.

4. Prepare a spreadsheet using spreadsheet software that will automatically calculate the diameter under the constraint of 2000 cm^3 for various depths.

5. After your data have been checked and validated, prepare a graph showing wind turbine diameter versus depth (or height). An example is shown in **Fig. 4-8**. (Note that terms such as "height" are dependent upon orientation. Since you will probably be building a wind turbine that operates on a horizontal axis, the "height" of a cylinder in these examples is actually the "depth" of the wind turbine.)

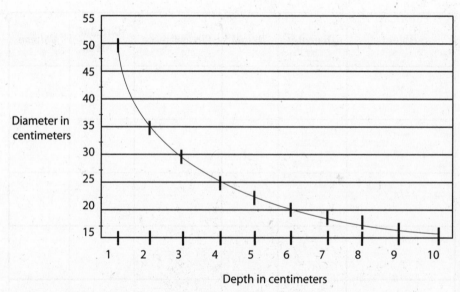

Fig. 4-8 Graph of diameter versus depth for a constant volume of 2000 cm^3

6. Use your graph to help you design your wind turbine so that it meets the design constraint of 2000 cm^3. For example, if you decide to build a collector with a diameter of 25 cm, then you can locate this point along the y-axis and determine that its depth cannot exceed about 4 cm. On the other hand, if you decide that the collector is to be about 3 cm in depth, then it cannot exceed about 29 cm in diameter.

THINKING CRITICALLY

1. Why is the line between the diameter and the depth of the wind turbine curved rather than straight? Can you think of other examples of relationships that would produce a curve?

_____

_____

_____

_____

_____

_____

_____

2. Can you think of other graphs that you could prepare that would be useful to you?

_____

_____

_____

3. For a fixed volume of 2000 cm^3, how are depth and diameter related?

_____

_____

_____

_____

WIND TURBINE DESIGN

Engineering Technology

For this part of the project, you will use the mathematics and science principles you have been working with to design your wind-powered generator that will convert wind energy into electrical energy. Several different blade designs are possible, such as curved, tapered, cup-type, and multiple pitch. **Figure 4-9** on page 104 shows one possible design.

NOTES

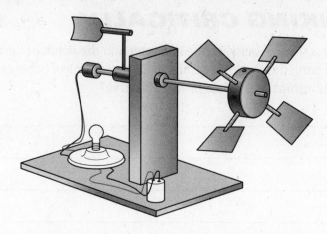

Fig. 4-9 Possible wind-powered generator design

MATERIALS AND EQUIPMENT

Quantity	Description
Assorted	illustrations of and reference materials on wind-powered devices
1	5-12 volt DC motor
1	light-emitting diode
1 30-cm length	insulated electrical hookup wire
2	multimeters (or 1 millivolt meter and 1 milliamp meter)
assorted	materials for propeller, vane, and support structure, such as wood, sheet metal, and sheet plastic
1	¼″ dia. x ¾″ long hex head cap screw
1 set	drill and numerically sized drill bits
assorted	fasteners and adhesives, including spot welding and soldering

Your teacher will show you a circuit similar to the one in **Fig. 4-10**. The same circuit will be used to test your wind-powered generator, except that the light-emitting diode (LED) will be replaced with a load resistor. The LED is used in this demonstration so that you will be able to see by its glow that electricity is being generated.

You must decide how to mount your turbine to the shaft of the motor/generator. A hole slightly smaller than the shaft of the motor could be drilled into the center of the turbine. Press the turbine onto the shaft. The recommended method, however, is to attach a cap screw to the shaft of the motor.

1. Clamp a ¼″ dia. × ¾″ long hex head cap screw in the chuck of a metal lathe with the drill bit mounted in a chuck in the tailstock. Drill the hole by advancing the bit manually into the bolt head while the bolt is spinning in the lathe. The hole should be slightly larger than the diameter of the generator shaft and close to the center of the head.

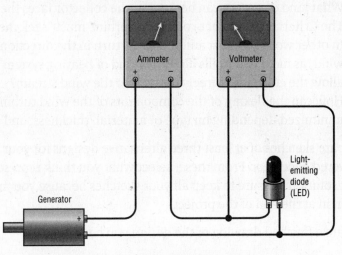

Fig. 4-10 Generator circuitry

2. Use epoxy adhesive to secure the cap screw to the shaft of the generator, as shown in **Fig. 4-11**. Keep the shaft pointed down while the epoxy cures so that it does not flow into the motor/generator.

3. Before you begin your design, review the Criteria and Constraints in the Design Brief. Also, consider the following factors in relation to the materials and processes to be used. For example:

   • Will the wind-powered generator rotate on a vertical axis or a horizontal axis?

   • Within the volume constraints, should the diameter of the wind turbine be maximized or should its depth be maximized?

   • The turbine's hub area restricts the flow of air. How can this dead area be minimized while maintaining a secure mount for the blades?

   • Should the shaft of the generator be mounted so that it faces toward the wind stream or away from it?

   • Should the mass of the wind turbine be kept at a minimum or can added mass be used to advantage as with a flywheel?

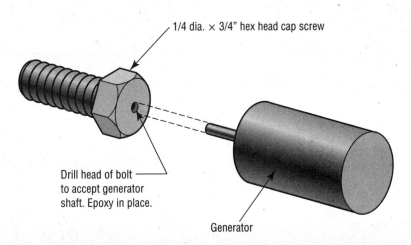

1/4 dia. × 3/4" hex head cap screw

Drill head of bolt to accept generator shaft. Epoxy in place.

Generator

Fig. 4-11 Mounting the wind turbine to the motor shaft

NOTES

- What kind of system can be used so the collector faces the wind?
- The Criteria specify that your wind turbine must "seek the wind." In other words, it must automatically turn in the direction of the wind, as most windmills do. What kind of bearing system can allow the collector to freely rotate into the wind stream?
- How can the flexing of the components of the wind turbine be minimized depending on type of material, thickness, and shape?

4. Prepare sketches of at least three alternative designs for your wind-powered generator. From these, select what you think represents the best solution. Be sure to keep all your sketches because you must turn them in at the end of the project.

5. Create a finished drawing of the design you have chosen.

THINKING CRITICALLY

If you wish, you may do research to find the answers to some of the following questions.

1. Where do you expect friction to occur in your device?

_____

_____

2. Should the blades of the turbine be light or heavy? Why? What might an advantage be to using heavier blades?

_____

_____

_____

_____

_____

_____

Modeling

During this part of the project, you will actually construct your wind-powered generator. Double-check the Criteria and Constraints to be sure your design meets them. Test, modify, and retest your solution until it provides maximum power output.

CONSTRUCTING A WIND-POWERED GENERATOR

Your teacher will set up the wind turbine testing apparatus so that you can measure the power output of your solution. A heavy duty, ceramic-bodied resistor, often called a power resistor, is used as a load for the generator. See **Fig. 4-12** on page 108. Your teacher will demonstrate how to measure the power output when you test your wind-powered generator. Note that you must take both voltage and amperage readings at the same time.

MATERIALS AND EQUIPMENT

Quantity	Description
assorted	materials for building the turbine propeller, vane, and support structure
assorted	fasteners and adhesives
2 or 3 per class	testing systems with safety shroud
2 or 3 per class	3-speed, 26″ electric fan
1	5-12 volt DC motor
2	multimeters (or 1 milliamp meter and 1 millivolt meter)
1	photo-tachometer (optional)

SAFETY FIRST

Before You Begin Make sure you understand how to use the tools and materials safely. Ask your teacher to demonstrate their proper use. Follow all safety rules and wear eye protection. Be especially careful when drilling thin sheet material. Refer to the Safety Handbook, beginning on page 291, for more information about safety in the lab.

Project 4 Modeling

NOTES

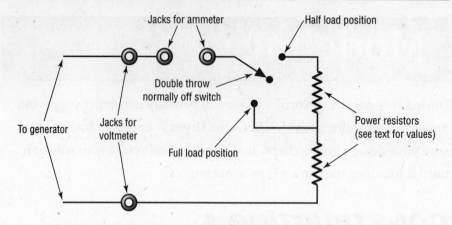

Fig. 4-12 Recommended circuit for determining the power output

1. Build your wind-powered generator according to the design you developed.

2. Test your device at one of the testing systems set up by your teacher. Be sure the safety shroud is in place.

3. Modify and retest your solution in order to optimize your turbine's performance.

4. Record your test results on the Test Data Table. On a separate sheet of paper, note the modifications you made each time.

5. After you have maximized your device's output, it is ready for testing.

TEST DATA TABLE

Test	Distance	Wind Direction	Potential Difference	Current	Power (V A)	RPM
1						
2						
3						
4						

THINKING CRITICALLY

1. If you found that heavier turbines performed better, how could you redesign your turbine?

   _____

2. If you found that your turbine was out of balance, what might you do to restore balance?

   _____

   _____

   _____

   _____

MEASURING ELECTRICAL POWER

Science

During this activity, you will use a voltmeter to measure potential difference and an ammeter to measure current in a generator circuit. You will then determine the power output from the generator in watts. This is done by multiplying the potential times the current. Later you will use these methods to calculate the output from your own wind-powered generator. The circuit used for testing is shown in **Fig. 4-13** on page 110.

SAFETY FIRST

Before You Begin Do not underestimate the danger of electrical shock in the lab or at home. Familiarize yourself with electrical safety as outlined in the Safety Handbook, beginning on page 291.

MATERIALS AND EQUIPMENT

Quantity	Description
assorted	lengths of insulated 22–26 gauge wire
1	hand-held generator
1	lamp to match generator voltage output
Assorted	light bulbs, resistors, and motors
1	ammeter
1	voltmeter

NOTES

The watt is a common unit used to measure electrical power. You will find wattage indicated on light bulbs and heaters, for example. (One watt is equal to one joule per second.) Because of the inefficiency of the generator used in this activity, the fairly low velocity of the wind stream, and the small size of the wind turbines, the power produced will be quite small, probably less than one watt.

1. Connect the circuit shown in Fig. 4-13. The ammeter is connected in series in the circuit since this is the only way in which the flow of the electrons can be measured. The circuit must be broken in order to insert the ammeter. The voltmeter, on the other hand, is connected in parallel with the circuit.

2. Operate the generator at two different speeds, one twice the other. The speed can be monitored by having one person count time intervals while another person turns the handle of the generator in sync.

3. Measure the voltage and current at the same time. Experiment with various loads, such as light bulbs, resistors, and motors.

4. Record the data for the various loads at the two different speeds in the Relationships Between Generator Speed and Power Table. To determine the power in watts, multiply the potential (voltage) times the current.

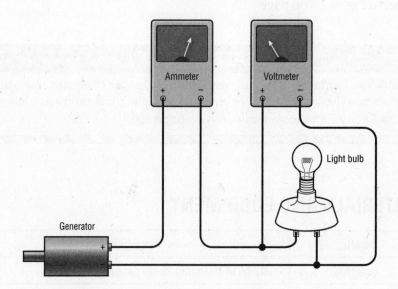

Fig. 4-13 Measuring voltage and current in generator circuit

Project 4 Modeling

RELATIONSHIPS BETWEEN GENERATOR SPEED AND POWER TABLE

Load	Speed	Voltage	Current	Power in Watts
	1 rev. per sec			
	2 rev. per sec			
	1 rev. per sec			
	2 rev. per sec			
	1 rev. per sec			
	2 rev. per sec			

5. What conclusions can you draw from these experiments?

_____

_____

_____

_____

THINKING CRITICALLY

1. Why is the handle of a generator more difficult to turn as the electrical load is increased?

_____

_____

2. Do ordinary higher wattage light bulbs have higher or lower resistance than lower wattage bulbs? Why?

_____

_____

_____

_____

Evaluation

During this part of the project, you will connect your wind-powered generator to a voltmeter and an ammeter. From the readings taken, you will determine its actual power output. You will also investigate the relationships among potential difference, current, and resistance.

TESTING YOUR WIND-POWERED GENERATOR

 During this activity, you will test your wind-powered generator in competition with the other generators designed by your classmates. Your device will be given six trials. Then an average of all the scores will be taken.

MATERIALS AND EQUIPMENT

Quantity	Description
1	generator testing apparatus
1	safety shroud
1	photo-tachometer (optional)

SAFETY FIRST

Before You Begin Make sure you understand how to use the tools and materials safely. Ask your teacher to demonstrate their proper use. Follow all safety rules and wear eye protection. Refer to the Safety Handbook, beginning on page 291, for more information about safety in the lab.

1. Be sure you are wearing eye protection during the testing procedure.

2. Connect your wind-powered generator to the circuit.

3. Locate the fan so that it is 15 cm (6″) from the nearest part of the wind turbine.

4. Rotate the base of the wind-powered generator (or change the position of the fan) to various positions while maintaining the 15 cm (6″) distance from the fan. If the collector meets the specifications for the challenge, it should automatically turn toward the wind source.

5. Take current and voltage readings six times and record them with other data in the Wind-Powered Generator Performance Record.

6. Total your device's scores and determine averages.

7. Increase the distance between the collector and the fan to 30 cm (12″) and collect the data at this new position.

8. Repeat data collection at 45 cm (18″). (Data from this additional distance will not be used in determining a winner but will be used for a later activity.)

WIND-POWERED GENERATOR PERFORMANCE RECORD

Trial	Distance	Wind Direction	Potential Difference	Current	Power (V A)	RPM
1						
2						
3						
4						
5						
6						
Total						
Average						

9. In design and engineering firms, design groups meet to go over performance data in order to make the best product. As a class, create a chart of information about all the turbine designs, including such things as design details, pitch, material used, and voltage produced by the two speeds of the fan. As a class, discuss these questions:
- What worked well?
- In general, how could the designs be improved?
- What changes, if any, would you make if you repeated this project?

Project 4 Evaluation

NOTES

THINKING CRITICALLY

1. What would happen to the speed of the wind turbine if the load resistor were disconnected?

_____

_____

_____

2. What is the relationship between the speed of the collector and the amount of potential produced?

_____

_____

3. How could you store the electricity for use when the wind is not blowing?

_____

_____

CALCULATING GENERATED POWER

For this activity you will learn more about the relationships among amperes, volts, and watts. Do you remember the symbols used when measuring electrical energy?

P = power in watts
I = current in amperes
V = potential difference in volts

Current can be determined using this formula:

current (I) = voltage (V) ÷ resistance (R)

Power is calculated using this formula:

$$P = I \times V$$

MATERIALS AND EQUIPMENT

Quantity	Description
assorted sheets	graph paper
1	generator testing apparatus
1	volunteer wind-powered generator
1	meter stick or meter tape measure

Project 4 Evaluation

1. Using the data from your Wind-Powered Generator Performance Record, calculate the power readings.

2. Study the data. Look at the voltage and amperage readings taken at each of the two distances. Predict what a graph of these data might look like and what information you might be able to observe.

_____

_____

3. Graph each of the readings on a single broken-line graph. The lines can be drawn in different colors. However, when a graph of this sort is made professionally, the data points are surrounded by a triangle for the voltage, a circle for the amperage, and a square for the wattage. This makes it easier to read the data points and also allows the graph to be photocopied. See **Fig. 4-14**.

4. If a photo-tachometer is available, you will have data on the speed of the collector in rpms at the various settings. Plot this speed on the graph versus the amount of voltage produced. There should be a direct relationship between these two variables.

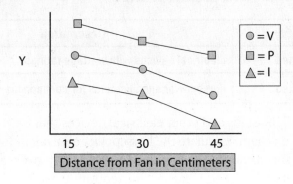

Fig. 4-14 Measuring voltage and current in generator circuit

5. Set up the testing apparatus and the wind turbine from a volunteer. With the fan located 6 cm from the wind turbine, take voltage and current readings and record them.

6. Increase the distance between the fan and collector in 2 cm increments. Do four readings of the current and voltage at each position.

7. After four readings have been taken, predict subsequent readings.

8. Take 16 more readings.

9. Prepare a line graph of distance vs. current and distance vs. voltage.

10. Predict the distance at which no power will be produced and then verify that distance.

Project 4 Evaluation

NOTES

THINKING CRITICALLY

1. Based on your graph, is the low speed of the fan half of the high speed? If not, what is it? Check the specifications on the fan used in the technology class and compare it to what your graph showed.

_____

2. How do different lines on the graph relate? What are the patterns?

_____

_____

CALCULATING HORSEPOWER

Horsepower is used to measure the power of cars and is also used to measure mechanical work. One horsepower = 746 watts. For this activity, you will become familiar with the power in one horsepower and one watt.

MATERIALS AND EQUIPMENT

Quantity	Description
assorted	electrical devices with wattage ratings
assorted	horsepower specifications for common machines

1. Using light bulbs and other electrical devices that have a wattage rating on them, calculate the horsepower equivalent for each device.
2. Using specifications for machines with a horsepower rating (e.g., electric motors), convert horsepower to watts.
3. Using data from the Wind-Powered Generator Performance Record, determine the horsepower of your wind-powered generator.

THINKING CRITICALLY

1. When you review ratings of horsepower for automobiles in magazines, how true do you think the values are in real applications?

_____

_____

2. What are some of the trade-offs in purchasing a dishwasher that has a ½ horsepower motor versus a ¼ horsepower motor?

_____

_____

_____

Electromagnetic Crane

Design Brief

When automobiles wear out, they end up in unsightly junkyards. Recycling has become more common in recent years, and nearly all parts of junked cars may now be recycled for profit. If enough recycling is done, many junkyards could be closed. First, however, the cars must be transported from the junkyards to the recycling centers. To do this, large electromagnetic cranes pick up the cars and place them in large transport vehicles, which then carry them away. See **Fig. 5-1**.

Fig. 5-1 This electromagnetic crane can be very effective for lifting heavy metal objects.

Challenge

Design and construct a working model of a crane-like electromagnetic device that could be used to remove cars from a community junkyard. It must be able to pick up five cars, lift them over a barrier, and then release them into a railroad car for transport and removal.

Criteria

- The cars will be represented by ¼" hex nuts and the railroad car by a tuna fish can.
- Your device must use an electromagnet to pick up each of the 5 cars, one at a time, lift it over a 15 cm high barrier that is 15 cm away, and release it into the railroad car.
- Your device must employ one or more cranks to control both the vertical and horizontal motion of the crane boom (arm).
- The device that moves all five cars to the railroad car in the shortest time will be considered the winner.

NOTES
- All designs must be accompanied by appropriate documentation, including:
  - ☐ Sketches of all possible solutions considered
  - ☐ A final drawing of your chosen solution
  - ☐ Data collected from various tests and experiments
  - ☐ A chart/graph showing how your solution performed
  - ☐ Information gathered from resources
  - ☐ Notes made along the way

Constraints

- You must be able to use your device to pick up and release the cars without touching the electrical connections or power supply.
- Your device must be able to pick up one, and only one, car at a time from a pile of other cars.

Engineering Design Process

1. **Define the problem.** Write a statement that describes the problem you are going to solve. The Design Brief and Challenge provide information.
2. **Brainstorm, research, and generate ideas.** The Research & Design section, beginning on page 117, provides background information and activities that will help with your research.
3. **Identify criteria and specify constraints.** Refer to the Criteria and Constraints listed on pages 115 and 116. Add any others that your teacher recommends.
4. **Develop and propose designs and choose among alternative solutions.** Remember to save all the proposed designs so that you can turn them in later.
5. **Implement the proposed solution.** Once you have chosen a solution, determine the processes you will use to make your electromagnetic crane. Gather the tools and materials you will need. Make sure you understand and follow all safety rules.
6. **Make a model or prototype.** The Modeling section, beginning on page 141, provides instructions for building the crane.
7. **Evaluate the solution and its consequences.** The Evaluation section, beginning on page 145, describes how to test the crane.
8. **Refine the design.** If instructed by your teacher, make changes to improve your design.
9. **Create the final design.** If instructed by your teacher, make a new crane based on the revised design.
10. **Communicate the processes and results.** Write a report about your project. Be sure to include all the documentation listed under Criteria.

Research & Design

During this part of the project, you will learn about concepts and tools that will enable you to design your solution. You will study simple machines and electricity. You will conduct an experiment in electromagnetism that relates directly to the solution you design and build. Also, you will work with the math relationships of simple machines and electromagnetism, as well as the geometry of cranes.

SIMPLE MACHINES

Science

The six simple machines are devices that create **mechanical advantage**. They include the lever, the wheel and axle, the pulley, the inclined plane, the wedge, and the screw. See **Fig. 5-2.** They make work easier by multiplying human force.

MATERIALS AND EQUIPMENT

Quantity	Description
1	5- or 10-speed bicycle
2 or more	pulleys
1	ball bearing

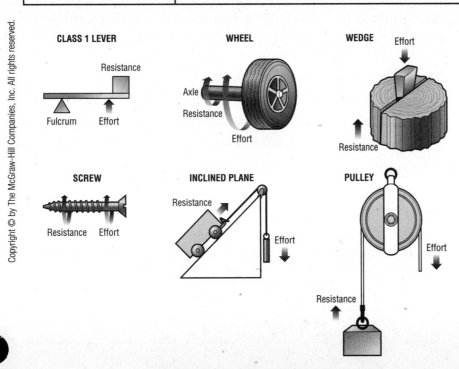

Fig. 5-2 The six simple machines

Name_____ Date_____ Class _____

Look at the crane shown in **Fig. 5-3**. Can you identify some of the applications of simple machines used in the crane? Can you think of ways simple machines could be used in your own electromagnetic crane?

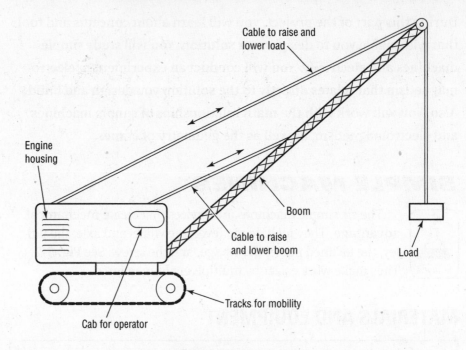

Fig. 5-3 Typical crane

Usually, mechanical power is changed in some way before it is put into use. Control devices like gears, pulleys, and sprockets and chains change its direction, force, or speed. See **Fig. 5-4**. Ball bearings are used to reduce friction at pivot axes. The weight of a crane arm, particularly when hoisting a car, is important, and ball bearings reduce the friction.

1. As a class, look at the bicycle's sprocket-and-chain system.

2. One student should lift the rear wheel off the floor, while another applies pressure to one of the pedals with one of his or her hands.

3. Change from one sprocket size to another. Does it take more or less force to move the pedal and wheel? How much force is required in relation to the speed of the wheel?

4. Look at the crane in Fig. 5-3 again. Would the engine pulley in such a crane be larger or smaller than the pulley around which the cables are wound?

_____

_____

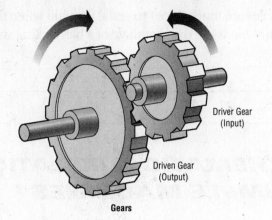

Driver Gear
(Input)

Driven Gear
(Output)

Gears

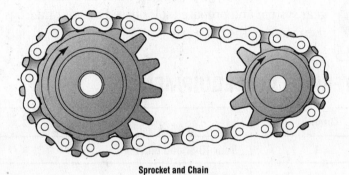

Sprocket and Chain

Fig. 5-4 Components that change direction, force, and speed

5. Look again at Fig. 5-3. Identify the rotational axes on the crane.

_____

_____

_____

THINKING CRITICALLY

1. What are some of the simple machines used in a crane?

_____

_____

_____

_____

_____

_____

_____

NOTES

2. Bicycles are much easier to pedal up a hill when they are in a low gear. What is the disadvantage of a bicycle that would operate only in low gear?

_____

_____

MATHEMATICAL RELATIONSHIPS IN SIMPLE MACHINES

Math

For this activity you will take measurements from the bicycle's gear system and prepare tables based on your findings. You will then analyze the data.

MATERIALS AND EQUIPMENT

Quantity	Description
1	5- or 10-speed bicycle
1	bicycle stand
7	weights, such as bricks
1	cloth tape measure
1 or more sets	Delta™ Gears (optional)

1. Measure the circumference of each of the sprockets on the bicycle. Record the data in the Bicycle Sprocket Analysis Table.

2. Count the number of teeth on each of the sprockets. Record the data in the Bicycle Sprocket Analysis Table.

3. While pedaling using your hand, set the bicycle for first (low) gear. Bring the wheel to a stop and adjust the pedals so they are parallel to the floor.

4. Place weights, such as bricks, on the pedals until the wheel starts to turn. The amount will vary from bike to bike. Record the approximate minimum weight needed to turn the wheel around.

5. Make a mark on the rear tire with chalk. Push a pedal slowly through one complete revolution and note where the white mark ends up. (You will need to restrict the wheel so it doesn't "freewheel" after the pedal has made one revolution).

6. Determine how far the bicycle would travel with each revolution of the pedal in this gear. Record this data on the Bicycle Crank Rotation Analysis Table.

BICYCLE SPROCKET ANALYSIS TABLE

Speed	Pedal Sprocket		Wheel Sprocket	
	Circumference	No. of Teeth	Circumference	No. of Teeth
1 (low gear)				
2				
3				
4				
5				
6				
7				
8				
9				
10 (high gear)				

Name_____ Date_____ Class _____

BICYCLE CRANK ROTATION ANALYSIS TABLE

Gear	No. of Crank Rotations	Distance Traveled
1 (low gear)		
2		
3		
4		
5		
6		
7		
8		
9		
10 (high gear)		

7. Shift into fifth gear, repeat Steps 5 and 6, and record the data.

8. Shift the bicycle into tenth gear, repeat Steps 5 and 6, and record the data.

9. Represent the data in the table visually with graphs, T-tables, and other means.

10. What conclusions can you draw from this experiment? Are the variables directly or inversely related?

_____

_____

_____

_____

NOTES

11. If sets of Delta Gears™ are available, conduct the same type of experiment using two different sized gears that are connected. Change out one of the gears and repeat the experiment several times. You may then set up different combinations of two gears and determine distance traveled. You may also add a third, fourth, and fifth gear to form a gear train. If you do this, be sure to note the direction (clockwise/counterclockwise) of gear movement in each case. Also note diameter and distance of travel for the final gear in the train during one revolution of the first gear. Record your data, represent them visually, and discuss your conclusions.

THINKING CRITICALLY

1. Develop a general rule for predicting the distance traveled based on the number of rotations of the pedal.

_____

_____

2. Suppose you conducted the same experiment with several different bikes. What variables do you predict would be the same on all bikes? Which variables would differ? How would this impact performance?

_____

_____

_____

_____

_____

_____

_____

_____

_____

NOTES

ELECTRICITY AND SIMPLE CIRCUITS

Science

Have you ever gotten a slight shock when touching an object? Everyone has experienced static electricity at one time or another. Did you know that lightning is a form of static electricity—its most dangerous form? In this activity, you will learn more about static electricity.

MATERIALS AND EQUIPMENT

Quantity	Description
2	glass rods, approx. 25 cm × 1 cm dia.
25 cm	ordinary sewing thread
2	plastic rods, approx 25 cm × 1 cm dia.
1	silk cloth
1	wool cloth
1	ring stand
2	clamps
1	DC variable power supply, 12 V max., 4 amps, breaker protected, with leads
2	1.5 V dry cells
1	parallel battery clip
20 feet	20-26 gauge insulated electrical wire
1	lamp with socket
1	5 V electrical switch

SAFETY FIRST

Before You Begin Make sure you understand how to use the tools and materials safely. Ask your teacher to demonstrate their proper use. Never leave unattended electromagnets or power supplies connected. Follow all safety rules and wear eye protection. Refer to the Safety Handbook, beginning on page 291, for more information about safety in the lab.

1. Hang two glass rods by a thread from a ring stand.

2. Rub both rods with a piece of silk cloth.

3. Bring the two rods close to one another from different directions. Describe what happens.

_____

_____

4. Repeat the experiment using the plastic rods and wool cloth. Describe what happens. Is there a pattern to the movement of the rods?

_____

_____

5. Why did the rods behave this way?

_____

_____

NOTES

The "spark" in static electricity is caused by the movement of electrons from atom to atom along a conductor. Rubbing the rods with a cloth causes them to build up electrons, giving them a negative charge. Objects with the same charge repel each other. If the rods are moved close to an object with a positive charge, they will be attracted to it. When the objects touch, the extra electrons escape.

What do you think is the difference between static electricity and the electrical current that powers the lights in the classroom?

_____

_____

6. Set up a simple circuit as shown in **Fig. 5-5**. Try different connections to learn what happens. What would happen if the wire used was of a different material? What if the wire was of a different gauge? What if the cells had higher voltage?

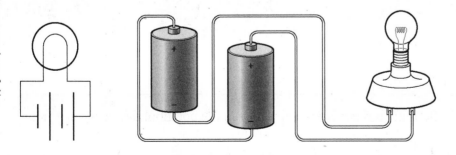

Fig. 5-5 **Electrical circuit**

NOTES

Here are some common terms used when studying electricity. Can you identify their relationships?

Term	Definition
Potential difference	A change in potential energy from one part of an electrical system to another. This creates the pressure, or voltage, that pushes an electrical charge through a circuit.
Current	The rate at which charges flow through a conductor. It is measured in amperes.
Resistance	A measure of how strongly a material resists the flow of electrons through it. It is measured in ohms.
Ohm's Law	A description of the ideal relationships among potential difference, current, and resistance: $E = I \times R$
Conductor	A material that allows electricity to flow through it
Insulator	A material that resists the flow of electricity

THINKING CRITICALLY

1. What are some of the ways in which electrical energy is controlled easily and safely?

   _____

   _____

   _____

   _____

2. What is a simple way to make the lamp in the circuit brighter?

   _____

   _____

   _____

3. What is the relationship between the potential difference in a circuit and the amount of work you want the circuit to do?

   _____

   _____

4. In a simple circuit, the potential difference across a light bulb was measured at 12 volts, and the current through the bulb was measured at 2 amperes. What is the resistance of the bulb?

_____

5. Why is it important to conserve electricity?

_____

_____

_____

N O T E S

MAGNETISM

Did you know that magnetism and electricity are related? In these experiments, you will learn how.

MATERIALS AND EQUIPMENT

Quantity	Description
1	magnetic compass
1 piece	thin cardboard
1	multimeter
2	bar magnets
1 tablespoonful	iron filings

SAFETY FIRST

Before You Begin Make sure you understand how to use the tools and materials safely. Ask your teacher to demonstrate their proper use. Follow all safety rules and wear eye protection. Refer to the Safety Handbook, beginning on page 291, for more information about safety in the lab.

1. Place a bar magnet on a table and place a thin piece of cardboard over it. Sprinkle iron filings on the cardboard in the area of the magnet. Gently tap the edge of the cardboard. Sketch the arrangement of the iron filings in the space below.

NOTES

2. What causes this to happen?

_____

_____

3. Pass the compass close to one of the poles of the magnet, then the other pole. What happens?

_____

_____

_____

4. Place the opposite poles of the two magnets close to one another. Then place the like poles close to one another. What happens?

_____

_____

5. Using the circuit shown in Fig. 5-5, place the compass near the conducting wire. What happens?

_____

_____

_____

_____

THINKING CRITICALLY

1. How far out are the filings affected by the magnet?

_____

_____

2. Describe the pattern formed by the iron filings when a magnet is held near them.

_____

_____

3. Where is the magnetic field concentrated around a magnet?

_____

_____

ELECTROMAGNETISM

During this activity, you will formulate and test hypotheses about the variables that affect the strength of an electromagnet. Your teacher will divide the class into teams and assign each team a different set of guidelines for making an electromagnet.

MATERIALS AND EQUIPMENT

Quantity	Description
1	DC variable power supply, 12 V max., 4 amps, breaker protected, with leads
Assorted	materials for electromagnet cores
Assorted	paper clips

SAFETY FIRST

Before You Begin Make sure you understand how to use the tools and materials safely. Ask your teacher to demonstrate their proper use. Never leave unattended electromagnets or power supplies connected. Follow all safety rules and wear eye protection. Refer to the Safety Handbook, beginning on page 291, for more information about safety in the lab.

1. Follow your teacher's guidelines for making your electromagnet.

2. Create a circuit. See **Fig. 5-6**. Connect a multimeter to your circuit to measure the potential difference and current. *The multimeter must be connected in series or it will be damaged.*

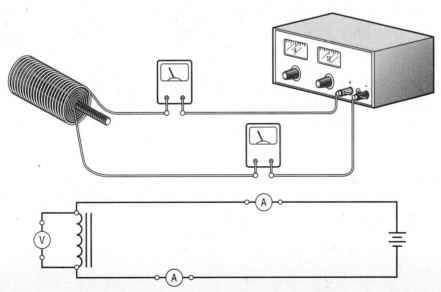

Fig. 5-6 Circuit including an electromagnet

NOTES

3. Using paper clips as a standard weight, determine how many clips your magnet can hold. Record the data in the Electromagnet Strength Table.

ELECTROMAGNET STRENGTH TABLE

Nunber of Turns	Core Material	Voltage	Current	Number of Clips Held

4. As a class, create a similar table on the chalkboard that everyone can see that shows all the teams' results.

5. As a class, discuss your experiments with electromagnets. Work out a formula for determining the power of an electromagnet.

_____

For the Challenge, you are required to build an electromagnetic crane that will pick up one, and only one, ¼ ˝ hex nut from a pile of them. How do you think you will apply what you've learned in this science activity to meeting this constraint?

THINKING CRITICALLY

1. What happens to a paper clip hanging from the electromagnet when the electricity is turned off?

_____

_____

2. What happens when you increase the number of loops or windings around the electromagnet's core?

_____

_____

3. Why would establishing a relationship between the number of windings and electromagnet strength be helpful? Couldn't you simply experiment to get the right strength?

_____

_____

_____

_____

4. It is possible that health risks are created by the electromagnetic fields around high voltage electrical transmission lines. Why is it difficult for scientists to determine the safety level of high voltage electrical wires?

NOTES

_____

_____

MATHEMATICAL RELATIONSHIPS IN ELECTROMAGNETS

During the previous experiment on electromagnetism, you completed the Electromagnet Strength Table. For this activity, you will use the data you collected to create graphs showing the various relationships.

MATERIALS AND EQUIPMENT

Quantity	Description
3 sheets	graph paper
1	computer and graphing software
several sheets	overhead transparency material
Several	marking pens

- Number of windings on the core compared to maximum weight lifted while the voltage was held constant
- Voltage compared to maximum weight lifted (using a variable power supply) while the number of windings was held constant
- Weight of core material compared to maximum weight lifted while the voltage was held constant

1. Using the data you collected and computer graphing software, graph the relationships you found. See **Fig. 5-7**.

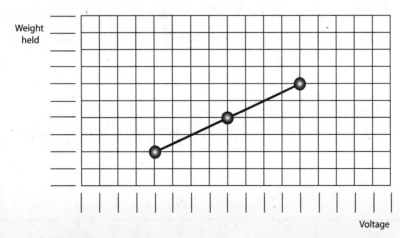

Fig. 5-7 Graphing the relationship between strength and voltage

NOTES

2. With your teammates, discuss the trends shown in your graphs. What types of relationships are present?

3. Develop an equation or mathematical model that describes the relationship you see in your graph of voltage versus weight held when windings are constant.

4. Briefly present your graph to the rest of the class. Prepare overhead transparencies and be prepared to communicate what your data and graphs suggest. For example, using the data shown in Fig. 5-6, if the voltage is "x" and the weight lifted is "y", then the slope of this line would be ½. This suggests that for every additional volt supplied, the weight that the magnet can lift will increase by two. To what extent can you infer such relationships using the data you collected?

THINKING CRITICALLY

1. What other things can we learn from a linear graph that are not shown directly on the graph?

_____

_____

2. Suppose you plotted a graph showing the number of windings compared to the weight the electromagnet held and found that the points plotted were close to linear but were not perfectly linear. What would you conclude from this?

_____

_____

_____

3. Suppose the plot line of a graph showing voltage compared to the weight the electromagnet held was extended backward until it crossed the y-axis. What meaning would this point have algebraically?

_____

_____

_____

4. How would a graph showing the relationship of differing core materials to the weight picked up by the electromagnet look? That is, how might it be similar or different from some of the other graphs produced by the class?

_____

_____

_____

Project 5 Research & Design

5. Describe the relationships among current, electromagnetic strength, and varying voltage. Try to generate mathematical models or equations about these relationships.

_____

_____

CRANE GEOMETRY

For this activity, you will use a model crane (see **Fig. 5-8**) to determine how crane size and motions affect distances traveled by the objects being hoisted. You will also construct tables and graphs based on the data.

MATERIALS AND EQUIPMENT

Quantity	Description
1	model crane
Assorted	pulleys
Assorted	lengths of string
1	chalkboard
1	light source for shadow drawing
3	soda straws
1	protractor

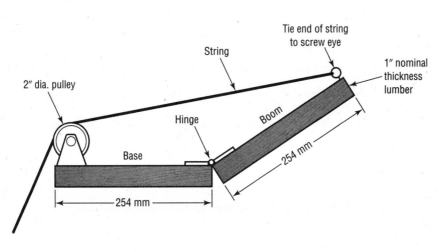

Fig. 5-8 Model of a simple crane

NOTES

1. Hold the model crane between a light source and a chalkboard so that its shadow is cast on the chalkboard. Place the crane boom in a horizontal position. As the crane boom is raised, trace the path of the tip of the boom on the chalkboard, much like a shadow figure. The tracing will be an arc. See **Fig. 5-9**. Where is the center point of the circle of which the arc is a part?

_____

2. On the board, draw the boom in the position to which it was raised.

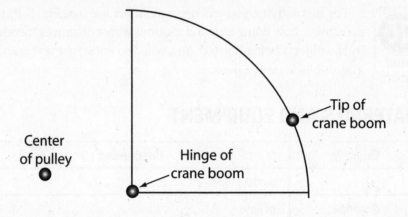

Fig. 5-9 Chalkboard tracing showing the path of the crane boom's tip

Then draw in the boom when it is horizontal. A sector of a circle is formed. What is the measure of its central angle?

_____

3. What fractional portion of one complete revolution (360 degrees) is this angle?

_____

4. What is the radius of the circle that contains this arc?

_____

5. What is the circumference of the circle with this radius?

_____

6. What fractional portion of the circumference is this arc length?

_____

Name________________________ Date______________ Class ______________

7. What is the relationship between arc length and circumference as compared to central angle measurement?

____

____

____

____

NOTES

Use the data you have collected to complete the Model Crane Data Table. Raise the boom to different heights. Add that data to the table.

MODEL CRANE DATA

Radius (crane boom length)	Description	Central Angle of Circle/ Arc	Circumference	Number of Clips Held
254 mm	Full circle	360°	20π	πd or $2r\pi$
254 mm	¼ circle			
254 mm				

Height of the Boom

What is the maximum height to which a boom can lift an object? At what angle with the base does this occur? Let's find out.

1. Construct a T-table showing angle and height of the boom tip. Prepare a graph of the data you collect. See **Fig. 5-10**.

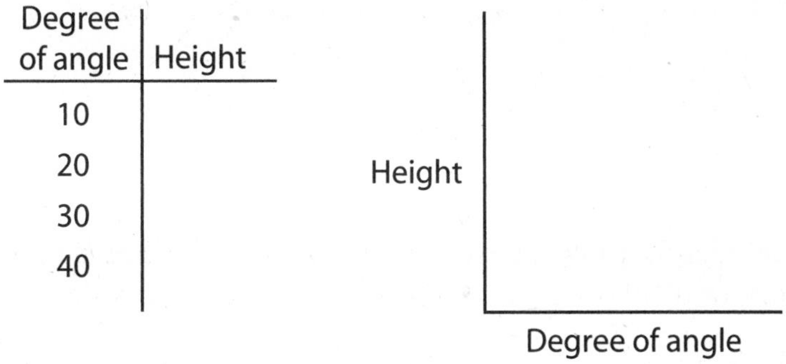
Degree of angle
Height
10
20
30
40
Height
Degree of angle

Fig. 5-10 T-table and graph

NOTES

2. What height does the tip reach when the boom is lifted 30 degrees from the base? Try other inclinations. What is the relationship between angle of inclination and height of tip?

_____

_____

_____

3. What angle would be necessary for the tip of the boom to be 5″ high? Try other heights. This is an example of interpolating data on a graph—constructing points between those for which you have data.

_____

4. Suppose you have three crane booms, one 171, one 254, and one 330 mm long. See **Fig. 5-11**. If each is raised to an angle of inclination of 30°, how high will the tip of each boom be? How is boom length related to height lifted for a constant angle?

_____

_____

_____

_____

_____

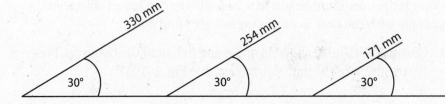

Fig. 5-11 Crane booms of different lengths

5. Cut the soda straws in lengths of 171, 254, and 330 mm, and use a protractor to determine degree of rotation. Place them on grid paper to look for relationships in question 8. See **Fig. 5-12**.

6. Consider the same three booms. Suppose you lifted the tips 127 mm. How would the angles of inclination of the three arms compare? To find their height, you will first need to construct a similar triangle. See **Fig. 5-13**. In geometry "similarity" is a term describing figures that are the same shape but are not necessarily the same size or in the same position. For the same angle, the longer the boom, the higher the boom tip. In this case, the boom length infuences height.

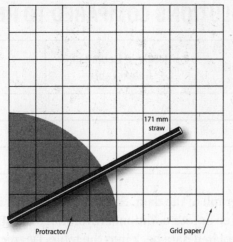

Fig. 5-12 Representing degrees of rotation

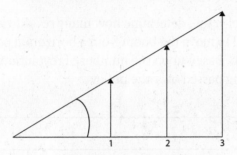

Fig. 5-13 Similar triangles

7. Is it possible for the crane in triangle 1 to lift anything as high as the one in triangle 3 in this sketch? Why or why not? Which two factors determine how high a crane lifts?

_____

_____

8. If three cranes with angles like those in Fig. 5-13 had the same cranking mechanism, which crane would be cranked to the proper height first? What implications does this have on the speed that the crane can complete its work of lifting five cars as required in the Challenge?

_____

_____

_____

Pulley Revolutions

Pulley revolutions also affect height of the boom. How are they related?

1. Raise and lower the boom while concentrating on the revolutions of the pulley wheel as compared to the height of the arm above the table surface. Enter these data into the Pulley Revolutions Compared to Height table for each of three revolutions of the pulley.

NOTES

PULLEY REVOLUTIONS COMPARED TO HEIGHT

No. of Revolutions	Height of Boom End Above Table	Degrees

2. Draw a line graph of the data. From the line graph, interpolate the height for three or four additional revolutions of the pulley. Then enter the interpolated data back into the table. Degrees (column three) may be included as an additional variable, if desired.

3. Using the model crane, determine how many revolutions of the pulley are required to move the boom from a horizontal position to a vertical position. Based upon the number of revolutions, determine how much string passed over the pulley.

_____

_____

_____

_____

_____

4. What would happen if the pulley on the model crane was replaced with one that is larger or smaller?

_____

THINKING CRITICALLY

1. If the boom angle is 45°, what is the height of the boom tip? If the height is 152 mm, what is the angle?

_____

_____

2. Compare and contrast this activity with the experiments you did with bicycle sprockets.

_____

_____

DESIGNING YOUR ELECTROMAGNETIC CRANE

Engineering Technology

During this part of the project, you will apply technology, science, and mathematics principles and techniques as you design your electromagnetic crane. One possible design is shown in **Fig. 5-14**. You will prepare sketches and a finished three-view drawing. Be sure to review the information in the Design Brief before you begin.

MATERIALS AND EQUIPMENT

Quantity	Description
Assorted	pictures of and resource material about cranes
several sheets	paper
1	pencil

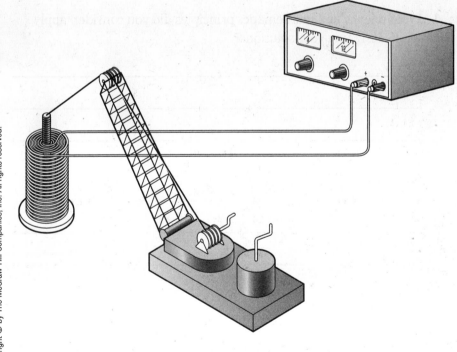

Fig. 5-14 Possible design for an electromagnetic crane

Name_____ Date_____ Class _____

NOTES

You have learned about the use of orthographic projection to create three-view, or working, drawings. These drawings show the most important views of an object, usually the front, top, and right side. Because an electromagnetic crane could be complex, limit your initial drawings to sketches. As you work, your designs will probably change, and you can edit your sketches more easily than a finished drawing.

1. Sketch top, front, and right side views for at least three different crane designs. Consider and apply the principles and methods you have been studying for this project.

2. If you wish, you may do research and use outside resources as you develop your designs.

3. From your sketches, select what you think represents the best solution. Be sure to keep all your sketches because you must turn them in at the end of the project.

4. Review the Criteria and Constraints. Does your chosen design meet them all?

5. Create a finished three-view drawing of the design you have chosen.

THINKING CRITICALLY

1. What science and mathematics principles did you consider/apply while designing your solutions?

_____

_____

_____

2. How did those principles influence your design?

_____

_____

_____

Modeling

During this part of the project, you will have an opportunity to use a wide variety of tools, materials, and procedures as you create your electromagnetic crane.

CONSTRUCTING YOUR ELECTROMAGNETIC CRANE

Engineering Technology

During this activity, you will construct your electromagnetic crane. Before you begin, be sure your device meets the Criteria and Constraints.

MATERIALS AND EQUIPMENT

Quantity	Description
1	DC variable power supply, 12 V max., 4 amps, breaker protected, with leads
1	¼″ × 2″ machine bolt and other core materials
10 m length	22-26 gauge electrical wire, varnish or enamel insulated
1 roll	PVC electrical tape
4	¼″ machine nuts
Assorted	light steel fasteners, materials and hardware
1	stopwatch or other timing device
Assorted	electrical connectors
2	power supply leads (connectors)
1 roll	thin rosin core solder
2 tubes	quick-setting epoxy adhesive
1	soldering pencil
1	wire wrapping jig
1	14 V switching device

SAFETY FIRST

Before You Begin Make sure you understand how to use the tools and materials safely. Ask your teacher to demonstrate their proper use. Follow all safety rules and wear eye protection. Refer to the Safety Handbook on page 291 for more information about safety in the lab.

Project 5 Modeling

NOTES

1. With your teammate, construct the electromagnet. See **Fig. 5-15**. Apply a layer or two of plastic electrical tape to the bolt so that the threads of the bolt will not damage the wire that will be wound around it later.

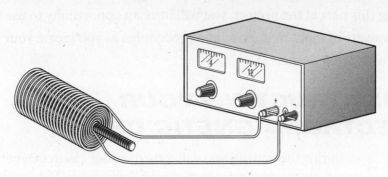

Fig. 5-15 Simple electromagnet with power supply

2. Attach the wire to the bolt just under the head with a short piece of tape. Leave about 60 cm (about 24″) of extra wire extending from the bolt so that it can be connected to the power supply after the coil is wound. The tape will hold the wire in place until a few windings have been made on the bolt.

3. Use the wire-wrapping jig to make the coil. See **Fig. 5-16**. Insert ⅜″ to ½″ of the threaded end of the bolt into the chuck of the wire-wrapping jig and wind the coil so that about one third of the wire has been wound around the bolt. *Avoid tangling the wires*. Count the number of windings that were applied and note this number. Use at least 100 windings. Cardboard washers can be added to the ends of the coil so that larger windings can be used.

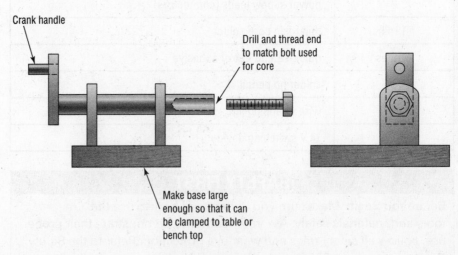

Crank handle

Drill and thread end to match bolt used for core

Make base large enough so that it can be clamped to table or bench top

Fig. 5-16 Wire-wrapping jig

4. Scrape insulation from the ends of the wire wherever connections are to be made.

5. Temporarily fasten alligator clips to the two leads extending from the coil, if necessary, to connect the coil to the power supply. Connect the coil to the power supply and determine if it will successfully pick up one and only one hex nut out of a group of hex nuts. Add or subtract windings from the coil so that it picks up only one nut out of the group, keeping track of how many windings are on the coil.

6. Once the coil is performing properly, cut the second lead so that it extends 60 cm (about 24″) from the bolt (the same length as the first lead). Neatly apply one layer of electrical tape to the coil and attach the alligator clips to the leads.

7. Consider mounting the electromagnet to the crane. See **Fig. 5-17**.

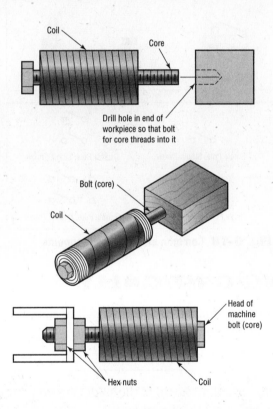

Fig. 5-17 Ideas for mounting the electromagnet

8. You have learned about pulleys, levers, ball bearings, and gears and your design should include any or all of these. How can common materials in the lab be used to make pulleys, levers, gears, and other components included in your design? See **Fig. 5-18** on page 144. Your three-view drawing may need to be reworked after you have experimented with various materials in the lab.

9. For safety, disconnect the electromagnet from the power supply as soon as you have conducted the necessary tests.

NOTES

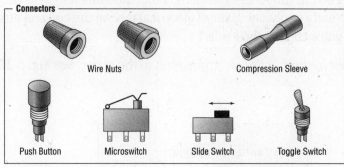

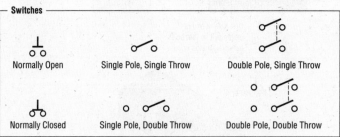

Fig. 5-18 Common electrical components

THINKING CRITICALLY

1. Why is varnish or enamel often used to insulate wire in coils and motors instead of plastic or rubber?

   _____

2. Why should you position the tip of the soldering iron and the solder on opposite sides of wire being soldered?

   _____

   _____

   _____

3. When using a wire nut, why should you twist the two wire ends clockwise?

   _____

   _____

Project 5 Modeling

Evaluation

During the Evaluation part of this project, you will test your solution to see how it performs. You will take the challenge two or more times. This will allow you to make modifications to your solution between runs. Data collected during the test will be analyzed and graphed.

TESTING YOUR ELECTROMAGNETIC CRANE

Engineering Technology

During this activity, your crane will be tested to see if it meets the challenge. Then you will evaluate it and make general observations about its performance.

MATERIALS AND EQUIPMENT

Quantity	Description
5 or more	¼″ hex nuts
1	empty, clean tuna fish can
1	1 × 6 board, 610 mm long
2	stopwatches or other timing devices

SAFETY FIRST

Before You Begin Make sure you understand how to use the tools and materials safely. Ask your teacher to demonstrate their proper use. Never leave unattended electromagnets or power supplies connected. Follow all safety rules and wear eye protection. Refer to the Safety Handbook on page 291 for more information about safety in the lab.

A pile of five or more ¼″ nuts to simulate old cars will be placed on one end of the "junkyard" site. The 1″ × 6″ board will be set up 15 cm from the pile of cars to simulate the fence marking the junkyard property line. On the opposite side of the fence and 15 cm from it, a tuna fish can be placed to simulate the waiting railroad car. See **Fig. 5-19**.

1. Place your crane wherever you like in relation to the pile of cars.

2. Record data on the Electromagnetic Crane Performance Record.

3. Two students will use the two stopwatches to measure your crane's performance time. One student should keep track of the total time, and the other should keep track of just the time taken to move each of the five cars (hex nuts).

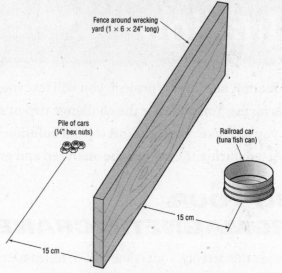

Fig. 5-19 Set-up for the performance test

ELECTROMAGNETIC CRANE PERFORMANCE RECORD

Team Name: _____

SPECIFICATIONS

Horizontal range in cm		Vertical range	
Operating voltage		Operating current	
Operating power			

TIME TRIALS

	Time in Seconds	Notes:
Auto #1		
Auto #2		
Auto #3		
Auto #4		
Auto #5		
Calculated Total Time		
Measured Total Time		
Average Time		

Project 5 Evaluation

THINKING CRITICALLY

1. To what extent does manual skill determine which crane performs the best?

_____

_____

2. Identify three uses of electromagnetism in your home.

_____

_____

ANALYSIS OF TEST RESULTS

During this activity you will construct and interpret tables, graphs, and rating systems to evaluate the electromagnetic cranes. You will be referring to your Electromagnetic Crane Performance Record.

MATERIALS AND EQUIPMENT

Quantity	Description
1	computer and graphic software

The word *best* has different meanings to different people and is usually based upon their own needs and values. Rating systems such as those in consumer product buying guides help to evaluate the overall effectiveness or value of a product.

Would adding up the points be a good system? Why or why not? One version is shown below. Do you think it works?

RATING BASED ON POINTS

Variable	5 pts.	4 pts.	3 pts.	2 pts.	1 pt.	0 pts.
Time required in minutes	< 1	1-2	2-3	3-4	4-5	> 5
Power in watts	< 5	5-10	10-15	15-20	20-25	25-30

Project 5 Evaluation

RATING BASED ON ORDINAL NUMBERS

Group	Time	Horizontal Range	Vertical Range	Power Required	Total Points
A	2	4	2	1	9
B	3	3	4	1	11
C	1	2	2	3	8
D	4	1	1	4	10

With the ordinal number system, for each variable, the cranes are ranked in order of best performance. For example, lowest power usage gets a 1 and highest vertical range gets a 1. If two groups score the same raw points, they both get the higher ordinal ranking (both are 1s or 2s). Then the next ordinal is skipped. In the column for Power Required, the values 1, 1, 3, 4 appear. This means that two groups had identical top scores. There were two firsts and no second. In the column for Vertical Range, two cranes rated with 2s. Thus, there were two second places but no third.

This type of ordinal rating system is used in figure skating to decide ties. If skaters have the same number of raw points, then the first, second, and third place ordinal rankings are added up and the lowest number wins.

To develop a rating system, the variables that could be a factor in determining "best" must be determined first.

1. Enter your crane's performance data into the software program.

2. As a class, discuss various ways of identifying the "best" crane. Ask:
   - Which crane would you purchase if speed were of greatest importance? What circumstances might demand a speedy crane?
   - Which crane would your company purchase if it had to pick up extremely heavy loads? Why?
   - What if you wanted the most flexible crane? Which one could be used for the greatest number of tasks? Why?

3. Rate your crane based on the points system. (In order for the crane to fit into a particular point category, both specifications must be met.)

_____

THINKING CRITICALLY

1. What kind of information is lost when a ranking system based on ordinal numbers is used?

_____

_____

2. Do you think the rating system chosen by the class was a fair one? How could it be improved?

_____

Rocket

Design Brief

Launching rockets is an expensive business. To offset the costs, NASA often launches *payloads*—cargo that produces income—into space for private companies. Communication satellites, for example, are "boosted" into space by NASA rockets. See **Fig. 6-1**. At the present time, a booster rocket falls back to Earth after delivering the payload and is destroyed in the process. However, if a booster rocket could return to Earth without being damaged, it might be reused, resulting in tremendous savings of money and time.

Fig. 6-1 This rocket is "boosting" a communications satellite into space.

Challenge

Design and build a rocket that will attain the highest possible altitude and return to Earth undamaged.

Criteria

- The rocket design must include some means of slowing the rocket's descent after it reaches maximum altitude.
- The rocket must be designed for a safe launch and a damage-free touchdown.
- The rocket must be tested for flight stability in a pre-launch test.
- A tracking device and two stopwatches will be used to track and time the rocket (1) from launch to maximum altitude (apogee) and (2) from deployment of the recovery system to rocket touchdown. The rocket's flight must be seen easily from the ground so that tracking is possible.
- Overall performance of each rocket will be determined by multiplying the maximum altitude reached by the "hang" time (duration of descent).
- You must document your work in a portfolio that includes the following:
    - ☐ Information gathered from resources
    - ☐ Three-view drawings of all the possible solutions you considered

☐ Charts/graphs/tables showing how your solution performed
☐ Notes made along the way

Constraints

- The rocket must be made of soft materials and have no metal parts, except for engine spacers, retainers, or wire used in the recovery system.
- The engine must be factory-made, factory-loaded, safety certified by the National Association of Rocketry (NAR), and correctly installed in the rocket body.
- The rocket must weigh no more than that specified for the engine that is being used.
- The recovery system must fit inside the rocket's nose cone.
- Materials used for the recovery system must *not* support combustion.
- The rocket must pass stability testing.
- The rocket must carry only the assigned payload.

Engineering Design Process

1. **Define the problem.** Write a statement that describes the problem you are going to solve. The Design Brief and Challenge provide information.
2. **Brainstorm, research, and generate ideas.** The Research & Design section, beginning on page 151, provides background information and activities that will help with your research.
3. **Identify criteria and specify constraints.** Refer to the Criteria and Constraints listed on pages 149 and 150. Add any others that your teacher recommends.
4. **Develop and propose designs and choose among alternative solutions.** Remember to save all the proposed designs so that you can turn them in later.
5. **Implement the proposed solution.** Once you have chosen a solution, determine the processes you will use to build your rocket. Then gather the tools and materials you will need. Make sure you understand and follow all safety rules.
6. **Make a model or prototype.** The Modeling section, beginning on page 174, provides instructions for building the rocket and the tracking device.
7. **Evaluate the solution and its consequences.** The Evaluation section, beginning on page 192, describes how to launch the rocket, track its position, and analyze the rocket's performance.
8. **Refine the design.** If instructed by your teacher, make changes to improve your design.
9. **Create the final design.** If instructed by your teacher, make a new or improved rocket based on the revised design.
10. **Communicate the processes and results.** Be sure to include all the documentation listed under Criteria.

At the moment the balloon was released, it moved very quickly and gained velocity. Thrust was greatest at that time and then decreased. Even after all the air was expelled from the balloon, momentum carried it forward.

Thrust works in a similar way in model rocket engines. Thrust changes over the duration of the flight. Most rocket engines exert a relatively powerful thrust (between 9 and 27 newtons, or 2 and 6 lbs.) during the first 5 seconds or so. Then the thrust levels off to a steadier but less powerful force (from 4 to 9 newtons, or 1 to 2 lbs.) for the next 3 to 10 seconds. Smaller engines produce a weaker thrust of much shorter duration.

The total change in momentum (total impulse) of the rocket can be determined by multiplying the thrust force by its duration in seconds.

THINKING CRITICALLY

1. How will the weight of a rocket affect its performance?

_____

_____

_____

2. How would using a slingshot illustrate Newton's third law?

_____

_____

ROCKET DESIGN

Engineering Technology

Can you think of real-life rocketry applications? Launching telecommunication satellites, space exploration, and defense and surveillance technology are all examples. Take a look at **Fig. 6-3** on page 154, which shows the details of a model rocket engine and engine placement. Then look at the model rocket **fuselage** (body) and engine your teacher has provided. Compare the model to the illustration in Fig. 6-3.

MATERIALS AND EQUIPMENT

Quantity	Description
1	model rocket fuselage
1	model rocket engine
assorted	sketching materials
3 sheets per student	drafting paper, 11″ × 17″
1 set per student	drafting equipment or CAD software

Name_____ Date_____ Class _____

ROCKET

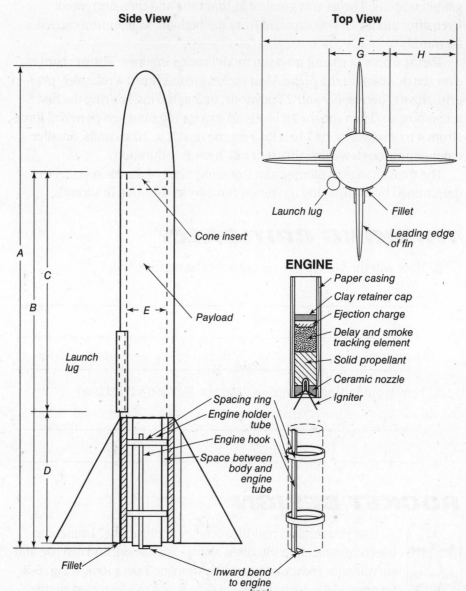

Fig. 6-3 This is a typical model rocket.

Now look at **Fig. 6-4**, which shows the basic rocket design you will be using for this project. Its components are described below, including design requirements and any areas in which you have freedom of design.

Rocket Tube

The diameter of the rocket tube is dependent on the diameter of the engine that is to be used. You may determine what length you want the tube to be. All tubes will be constructed by wrapping paper around a mandrel (rod) and then covering the paper with wrappings of gummed tape. You have your choice of finishes for your rocket tube.

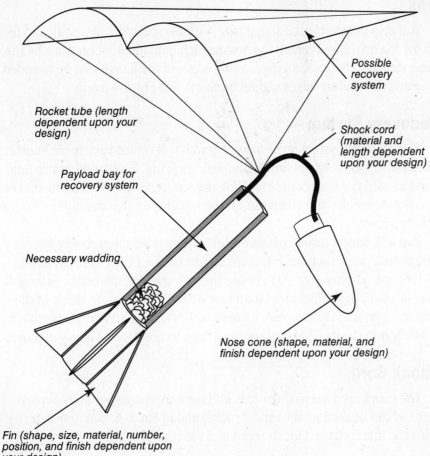

Possible recovery system

Rocket tube (length dependent upon your design)

Shock cord (material and length dependent upon your design)

Payload bay for recovery system

Necessary wadding

Nose cone (shape, material, and finish dependent upon your design)

Fin (shape, size, material, number, position, and finish dependent upon your design)

Fig. 6-4 This sample solution indicates areas in which you can exercise options when designing rocket. However, the engine size should be the same for every rocket, and the engine retaining and ignition techniques should not deviate from those specified by the engine manufacturer.

Payload Bay Area

The **payload** is the cargo, or load, that the rocket is intended to transport. The payload carried by a NASA rocket, for example, might be a communications satellite. The payload for your rocket will be the recovery system. The payload bay area will hold the payload until the payload is deployed. The diameter and length of the bay will be determined by the diameter and length of the rocket tube.

Nose Cone

The nose cone may be made from cork, Styrofoam™, or any other material you select. You may shape the nose cone as desired, but keep in mind that the shape should reduce friction as it passes through the air. The length of the nose cone is also up to you. You can see from the rocket model how the nose cone is attached to the rocket tube. You may also choose the finish for the nose cones, which should be smooth as possible to reduce friction.

NOTES

Fins

You may determine the shape, size, number, position, material, and finish for the fins. However, all the fins on each individual rocket must be the same size and shape. Also, the leading edges of the fins should be rounded over and the trailing edges should be sharp. See Fig. 6-4 again.

Recovery System

The recovery system is the assembly that will return your spent booster rocket safely to Earth. As the engine finishes firing, it will discharge a final burst toward the nose cone, expelling the recovery system from the rocket. The slower the descent (hang time), the less chance that damage will occur at touchdown.

You will have a variety of options in the design of the recovery system. One option is a parachute, but consider alternatives such as streamers or glider-type systems. You may choose from a variety of materials, as long as they are both lightweight and flame-retardant. The size and shape of the recovery system may also vary a great deal, but the system must be able to fit easily into the payload bay area and then be quickly and easily deployed.

Shock Cord

The shock cord keeps the rocket and nose cone together after deployment. It can be seen on the model rocket and in Fig. 6-4. You may determine the material used for the cord and the cord's length.

Making Preliminary Designs

See the model rocket engine specifications in **Fig. 6-5**. They contain information that you may want to consider when designing your rocket and recovery system.

Engine Size	Average Thrust in Newtons	Total Impulse Range in Newton-seconds (Average thrust x thrust duration)	Altitude Range of 1-oz. Model in Meters	Approximate Altitude of Typical Model in Meters	Minimum Diameter of Launch Site in Meters
1/4A3-2	3	0.63	15-75	30	15
1/2A6-2	6	0.63-1.25	30-120	57	15
A8-3	8	1.25-2.50	60-195	110	30
B6-4	6	2.50-5.00	90-300	220	60
C6-5	6	5.00-10.00	105-450	300	120

Source: Estes Industries, Inc.

Fig. 6-5 Model rocket engine specifications

1. Sketch at least three design ideas for a rocket. Compare and discuss your design ideas with those of your teammate.
2. Select one of the six designs for the rocket. Keep in mind that the finished rocket should be as light in weight as possible.
3. Each team member should prepare full-size preliminary sketches of your team's rocket. Later you will make any needed changes.
4. Next, sketch several preliminary designs for your rocket booster recovery system. The recovery system must slide easily into and out of the rocket body tube, so size will be an important factor.

NOTES

THINKING CRITICALLY

1. What would be the disadvantage of using a stiff material in your recovery system, even though you could fold it to fit?

_____

2. Why should the trailing edge of a fin be sharp?

_____

DRAG FORCES AND STREAMLINING

Science

Drag is the friction resistance experienced by a body moving through a fluid medium. The body is then slowed down. (Remember that a fluid can be a gas, such as air, or a liquid, such as water.) The resistance is distributed over the entire surface of the object.

MATERIALS AND EQUIPMENT

Quantity	Description
1 per student	index card, 3″ × 5″
1 per student	index card, 5″ × 8″
several	fans
2	balls denser than water and about 5 cm in diameter
1	cube as wide as the ball's diameter and about equal to it in mass
2	transparent containers about 30 cm tall and twice as wide as the ball's diameter
enough to fill the containers	water
enough for the class	assorted sketching materials
1 bottle	corn syrup, vegetable oil, or similar liquid denser and/or more viscous than water
1 roll	masking tape

NOTES

1. Hold a 3" × 5" index card and then a 5" × 8" index card in front of a fan, so that the cards are perpendicular to the wind stream. Feel the relative amount of force that is applied to the two cards. Note your observations below.

_____

_____

2. Hold one of the cards in front of the fan again. This time, change the angle of the card. Instead of holding it perpendicular to the stream of air, hold the card parallel to it. Note your observations below.

_____

3. Another factor that affects drag is the shape of the object. As a class, fill the two transparent containers with water. One student should take one of the 5-cm balls and stand in front of one of the containers. Another student should take the cube and stand in front of the second container. The width of the cube is equal to the diameter of the ball, and both objects are equal in mass. Which object do you think will fall to the bottom of its container first? The two students should release their objects at the same time into the tall containers of water. Note your observations below.

_____

_____

As an object moves through a fluid, turbulence (disturbance) is created in the fluid. The smoother the object's surface, the less turbulence is created. **Streamlining** is a design technique used to make objects sleek and to reduce turbulence. See **Fig. 6-6**.

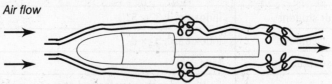

Air flow

Sharp angles create eddy currents and turbulence, thus increasing drag.

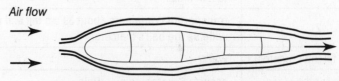

Air flow

Smooth, tapered surfaces reduce eddy currents and turbulence, thus reducing drag.

Fig. 6-6 The effects of streamlining

What objects can you think of that are streamlined? How do body shapes of automobiles from the 1930s and 1940s compare to those of today? Do you know what trucking companies do to their large trucks to reduce drag and increase fuel economy? As air passes over the top of the cab, it hits the semi-trailer unit like a wall. Trucking companies often mount airfoils, or "spoilers," on the roof of the cab to provide streamlined contours to the top of the truck to reduce wind drag.

4. As a class, attach tabs of masking tape to the surface of one of the identical 5-cm balls, as shown in **Fig. 6-7**, to give the ball a rough surface.

5. Refill the two transparent containers with water.

6. One student should take the taped ball and stand in front of one of the containers. Another student should take the smooth ball and stand in front of the second container.

The two students should release their balls at the same time into the tall containers of water. Note your observations below.

_____

_____

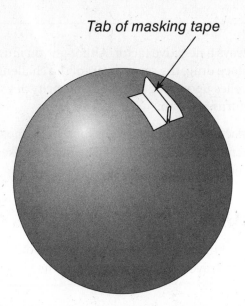

Tab of masking tape

Fig. 6-7 Apply masking tape to a ball, and drop the ball into a transparent container of water to demonstrate the effect of surface roughness on drag.

Think about how you can make the surface of your rocket as streamlined and smooth as possible in order to reduce drag.

Have you ever tried to "run" in a swimming pool? It was difficult because water is denser and more viscous (thicker) than air. Moisture-laden air that occurs in conditions of high humidity is denser than dry air. Denser air increases drag. In fact, pilots consider the density of the air as they plan flights.

NOTES

7. Replace the water in one of the transparent containers (used in previous experiments) with corn syrup, vegetable oil, or a similar transparent liquid that has a density and/or viscosity greater than that of water.

8. One student should drop a 5-cm ball into the container of water at the exact time that another student drops the other ball into the container of denser and/or more viscous liquid. Note observations below.

_____

Have you ever held a hand out of the window of a moving automobile? Did you notice that the drag on your hand was noticeably greater at higher speeds? You may have also noticed the relationship between velocity and drag when riding a bicycle. While pedaling at a slow speed, the drag you feel on your body is hardly noticeable. However, at higher speeds, the force is significant. Drag increases as the velocity of the moving object increases. Did you discover any way that you could reduce your pedaling effort while maintaining or even increasing your speed?

9. You may have tried crouching down over the handlebars. Why would this type of action help reduce drag?

_____

Drag is not always a negative factor. Although you must try to design your rocket to reduce drag, the second part of the challenge—using a recovery system—relies upon drag to limit the velocity at which your spent booster rocket returns to Earth.

10. Develop some new possible designs for the fins, nose cone, and recovery system of the rocket you will be making. Sketch these designs and place them in your portfolio.

THINKING CRITICALLY

1. Owners of pickup trucks drop their tailgates down to a horizontal position when traveling at high speeds on the highway. Why?

_____

_____

_____

2. How is streamlining evident in the bodies of sharks, porpoises, and other fast-swimming creatures?

_____

_____

WIND, MASS, AND ROCKET STABILITY

Science

Suppose a car is parked in a driveway, and a 20-km/hr wind is blowing. Molecules of air push against the car even though the car is at rest. If no wind is blowing but the car is driven down the road at 20 km/hr, the car will again encounter molecules of air, with the same force. See **Fig. 6-8**.

MATERIALS AND EQUIPMENT

Quantity	Description
1 per team	meter stick or yard stick
1 per team	small weights (e.g., coins, washers, machine nuts)
1 or more rolls	transparent or masking tape
1 sheet	Card stock at least 20-cm square
1	scissors
1	Frisbee® or similar disk toy
1	model rocket
1	ruler
1	felt-tip marking pen

Wind blowing at 20 km per hour   Car at rest

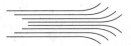

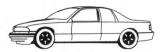

Is the same as:

No wind blowing   Car traveling at 20 km per hour

Fig. 6-8 Drag forces and relative wind

Project 6 Research & Design

NOTES

Whether the vehicle is moving or the air is moving, there is a stream of air affecting the vehicle. The movement of an air stream relative to the vehicle is called **relative wind** because this wind exists *relative to the vehicle*.

If no wind were blowing, a rocket in flight would follow a vertical trajectory (path). The relative wind force would come entirely from the rocket pushing against the atmosphere. See **Fig. 6-9**, part A. However, in real life natural wind is almost always present (part B). The natural wind combines with the wind caused by the rocket's motion to produce the resultant relative wind shown in part C.

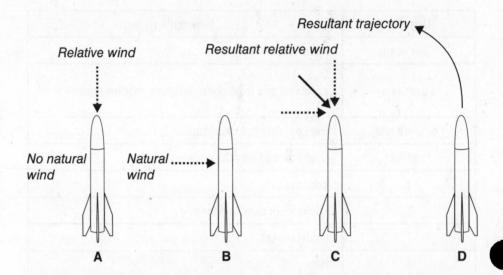

Fig. 6-9 Component of the relative wind on a rocket. In the absense of natural wind, the relative wind comes entirely from the force of air against the nose of the rocket as it pushes upward in A. The natural wind is shown in C. The stable rocket in flight tends to line itself up parallel to the relative wind, but it moves in the opposite direction to it in D.

Suppose a stable rocket launches on a day when a strong natural wind is blowing from left to right. How do you think it will affect the rocket's path? If you supposed the rocket would arc in the same direction the natural wind is blowing, you would be mistaken. In fact, the rocket will tend to turn slightly to face *into* the natural wind. See Fig. 6-9. Do you think you know why this happens? Let's find out.

1. Place a meter stick flat across your index fingers as shown in **Fig. 6-10**. Start by placing your fingers near the ends of the meter stick and gradually move them simultaneously toward each other.

2. Stop when your fingers meet in the middle of the meter stick and the stick remains steady. This point is the stick's **center of mass**—the point at which its mass is balanced.

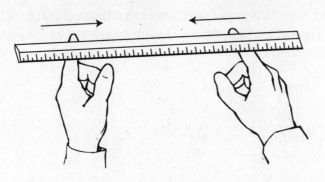

Fig. 6-10 Finding the center of mass of a meter stick

3. Tape a small weight, such as a coin, somewhere between the center and one end of the meter stick. Repeat steps 1 and 2. Is the center of mass still in the center of the meter stick?

_____

4. Using the card stock, cut a round disk that is about 20 cm in diameter. Tape a coin or other small weight to the center of the disk. Holding the disk by the edge, toss it into the air with a spinning motion. Observe how the disk moves.

5. Remove the coin and reposition it about 3 cm from the disk's center. Toss the disk into the air with a spinning motion. What do you see?

_____

_____

6. Predict what will happen to the center of mass of a Frisbee™-type disk if the coin is taped 3 cm from its center.

_____

_____

_____

_____

_____

_____

7. Tape the coin to the Frisbee-type disk and throw it into the air. Note your observations and conclusions.

_____

_____

Name_____ Date_____ Class _____

N O T E S The center of mass is one of two factors that help explain a rocket's trajectory. See **Fig. 6-11**. The second is the drag produced by the wind. They often work together.

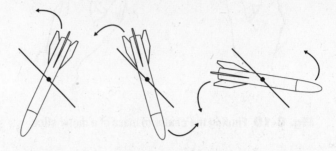

Fig. 6-11 When tossed in the air, the model rocket (or any object) rotates about its center of mass.

Since there is more surface area between the center of mass and the nose cone than there is between the center of mass and the tail of the rocket, the wind has more of an effect on the upper rocket body. This can cause the rocket to veer away from a course parallel to the relative wind and rotate around its center of mass. The rocket becomes unstable (and dangerous. See **Fig. 6-12**.

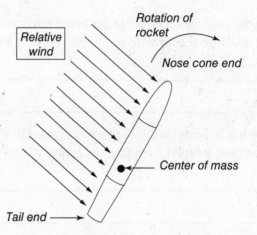

Fig. 6-12 When there are no fins on this rocket, the relative winds push more against the nose cone end of the rocket since it has a larger surface area. This causes the rocket to be unstable.

Fins can help compensate for this effect and can create a more stable rocket. Do you remember the experiment in which you held an index card in front of a fan? When the card was held perpendicular to the flow of air, the force of the wind pushed against the face of the card. If you had not held it steady, the wind would have pushed the card flat and parallel to the direction of the wind. When the wind encounters one or more fins, it applies its force against them. See **Fig. 6-13**. If the rocket is not parallel, it rotates around its center of mass, until the fins are parallel to the relative wind.

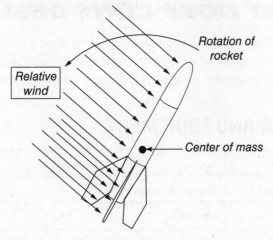

Fig. 6-13 The relative wind applies force to the fins, rotating the rocket about its center of mass and toward the relative wind.

THINKING CRITICALLY

1. Would huge fins maximize a rocket's stability? Why or why not?

_____

_____

2. How does adding fins to the tail affect center of mass?

_____

_____

3. How could you move center of mass closer to the nose of a rocket?

_____

_____

4. What would happen if the fins on a rocket were located near the nose instead of the tail?

_____

_____

5. Is a rocket more likely to veer away from a straight-line course shortly after launch or near the point of its apogee (highest point)?

_____

_____

_____

NOTES

ROCKET NOSE CONE DESIGN

For this activity, you will design and draw a template for a rocket nose cone. Do you know what a right circular cone looks like? In a right cone, the apex (top) is directly above the center of the base.

MATERIALS AND EQUIPMENT

Quantity	Description
1 large sheet	construction paper
1	compass
1 pair	scissors
several assorted	right conical shapes (1 of the same diameter as model rocket body; several with slightly smaller and slightly larger diameters)
1	protractor
several pieces	cardboard
1 piece about 200 cm3	modeling clay
1	dinner knife
1	funnel
1	model rocket body
assorted types	drawing materials
1	calculator

Nose Cone Geometry

What dimensions are needed to fully describe the size and shape of a cone? See **Fig. 6-14** on page 167. When the cone is opened and laid flat, it forms part of a circle. The circle's radius is equal to the side length of the cone. The ratio of the angle (β) between the two sides of the flattened cone to 360 degrees is proportional to the ratio of the circumference of the cone's base to the circumference of a circle with radius equal to the side length.

Suppose the cone in Fig. 6-14 is 4.0 cm high and has a radius of 1.2 cm. The circumference of the base of the cone is $2\pi r = 7.5$ cm, and the side length (s) can be obtained from the Pythagorean theorem: $s^2 = 4.0^2 + 1.2^2$. For this example, $s = 4.2$ cm (approximately), and the bottom edge will be an arc from a circle with a radius of s. The ratio of the measure of the angle β to 360 degrees is proportional to the ratio of the arc length (cone's bottom) to the circumference of a circle with radius s. Thus,

$$\frac{\beta}{360} = \frac{\text{arc length}}{2\pi s} = \frac{2\pi s}{2\pi s} \qquad \beta = \frac{(360)(1.2)}{4.2} = 103 \text{ degrees}$$

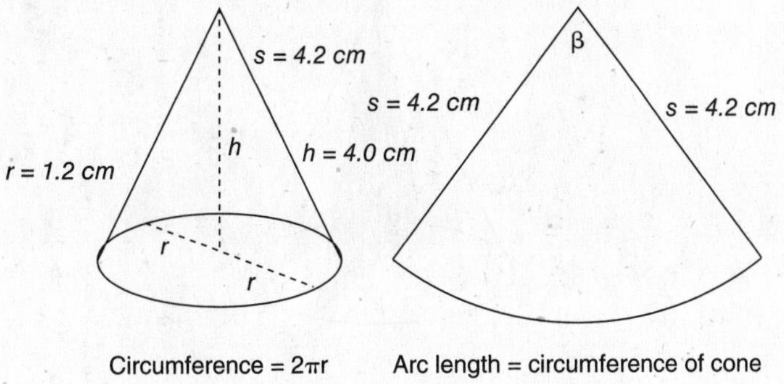

Fig. 6-14 Stretchout of a cone

1. Use a protractor to sketch this angle and draw its sides.

2. Then use a compass set to a radius of *s* to sketch in the arc. Add a "tab" on one side to be used in gluing the cone sides together.

3. Cut out the shape and form it into a cone.

The volume of a nose cone can also be a design consideration. Volume has a direct relationship to the weight of the nose cone for materials that have equal density. The formula for finding the volume of a cone is $V = \pi r^2 h / 3$. V and h are directly proportional if the radius, *r,* is constant for a particular rocket.

4. Construct a graph with V on the vertical axis and h on the horizontal while *r* is constant. The graph will form a straight line. This graph could be used to find the volume of a nose cone when a particular height is needed and vice versa.

Experimenting with Clay Shapes

Working with clay shapes can also help you design a better nose cone for a rocket.

1. Construct a right circular cone from modeling clay. If desired, you may use a funnel for a mold.

2. Refer to **Fig. 6-15** on page 168, and cut the cone to produce two identical pieces.

3. Place the cross-sectioned surfaces against a sheet of paper and trace their outlines. Describe the shape that you have produced below. Then label your tracing.

_____

4. Re-mold the clay back into a cone.

NOTES

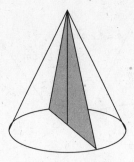

Fig. 6-15 Make a vertical cut through the clay cone to make two identical halves. The cross-sectioned surface is a triangle.

5. Slice through the clay cone parallel to the base. What shape have you produced?

_____

Trace the shape on a sheet of paper and label it.

6. Re-mold the clay into a right circular cone. Cut through the cone at an angle of roughly 15° to 30° to the base, but do not intersect the base. See **Fig. 6-16.**

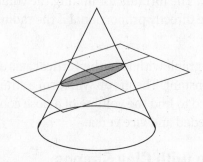

Fig. 6-16 What figure is created?

7. Place the cross section against a sheet of paper and trace its shape. Describe the shape below and label your tracing.

_____

_____

_____

8. Re-mold the clay into the shape of a cone. Cut the clay so that the cutting plane is parallel to the side of the cone as shown in **Fig. 6-17**. Hold the cross section against a sheet of paper and trace its outline. Label the tracing a *parabola*.

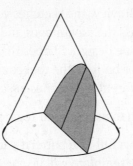

Fig. 6-17 Parabolic conic section

9. Re-mold the clay into a cone. Slice vertically through the clay starting at a point about halfway along one of the cone's sides. See **Fig. 6-18**. Place the cross section against a sheet of paper and trace the shape. Label the tracing a *hyperbola*.

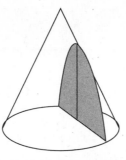

Fig. 6-18 Cutting the cone in this manner (starting halfway along one side and cutting straight down to the base) results in a hyperbolic cross section.

Designing the Nose Cone

You will now use what you have learned about right circular cones to design the nose cone for your rocket. Since the nose cone will fit on top of the rocket body, how should the diameter of the nose cone compare to the diameter of the rocket body? If you said that the diameter of the nose cone should be the same as the diameter of the fuselage, you are correct.

NOTES

If it is larger, it will increase drag on the rocket. If it is smaller, it cannot be mounted to the rocket body. There are also practical limitations on the height of the cone. A very high cone will be too fragile to be practical.

You and your teammate may wish to discuss other design-related questions such as: Does the diameter of the cone depend on the height? Theoretically, how many different nose cones could be designed for one particular rocket body? Review the sketches you made of the various conic cross sections, and discuss the advantages and disadvantages of each shape as the basis for a nose cone design. Shapes that provide a smooth transition between the nose and the body of the rocket are preferable.

1. Design several different nose cones based on different parameters.

2. Choose one design to use for your team's rocket.

3. Make a template—a full-size pattern—to use in the construction of the nose cone. The template will be a vertical cross section of the nose cone design.

4. Suppose that the rocket body is radius *r* and the height of the nose cone is *h*. What shape template needs to be constructed for the nose cone?

_____

_____

5. Measure the diameter of your rocket body. Using this measurement, draw a circle with the same diameter, and then cut out a semicircular template for a nose cone that will fit your rocket. See **Fig. 6-19**.

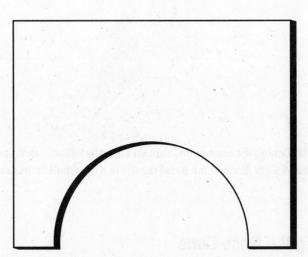

Fig. 6-19 A semicircular template for a nose cone

6. What is the longest nose cone you could construct using a semicircular template?

_____

_____

_____

NOTES

7. On a separate sheet of paper, write a paragraph explaining why you chose this particular nose shape and place it in your portfolio. Also include a drawing of the nose shape produced from the template and the sketches of the various conic designs you considered.

THINKING CRITICALLY

1. Would it be possible to use a single template for making nose cones for rockets with different body diameters? Why or why not?

_____

_____

2. Name some products that have nose shapes designed to reduce drag.

_____

_____

ROCKET FINS

Each rocket will have fins located around the bottom of its body. These fins are to be equally spaced. For this activity you will determine the angle between fins for your team's rocket design. Because you are free to choose the number of fins, the angle between the fins will vary from rocket to rocket. The rocket body is cylindrical, so its circumference will be 360 degrees. In general, if there are n fins, the angle between fins would be $360/n$ degrees.

MATERIALS AND EQUIPMENT

Quantity	Description
1	protractor
1	compass
1 sheet	graph paper

NOTES

Devise a plan to mark the rocket body at the appropriate places for the placement of the fins. One way would be to use a diagram of the circular base of the rocket body and mark the appropriate angles with a protractor, then transfer these measurements to the side of the rocket body. Another way would be to divide the circumference of the rocket body into *n* equal arcs where *n* is the number of fins desired.

Some fins may be shaped in the form of a complex polygon. The area of these fins could be easily computed by subdividing a full-size drawing of the fin into rectangles and/or triangles and computing the areas of these shapes.

If a fin is irregularly shaped, you must estimate its area. One method would be to draw the fin's outline on a piece of graph paper that has small

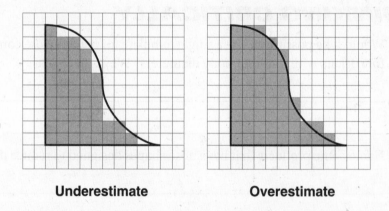

Underestimate   **Overestimate**

Fig. 6-20 Estimate the area of an irregularly shaped fin by taking a mean of an underestimate and an overestimate of the area the fin covers.

divisions and take the mean of an underestimate and an overestimate of the area the fin covers. See **Fig. 6-20**.

Even if you plan to use regularly shaped fins, it is useful to know how to under- and overestimate. Carry out the following steps using the example in Fig. 6-20.

1. Make an underestimate by counting the number of whole squares included within the outline of the fin in Fig. 6-20.

   _____

2. Make an overestimate by counting the number of whole and partial squares included in the outline of the fin. Some of the squares will extend beyond the fin's outline.

   _____

3. Take the mean (average) of the underestimate and the overestimate.

NOTES

_____

4. Draw the fin design for your own rocket on graph paper. If it is irregularly shaped, repeat Steps 1–3 to estimate its area. Otherwise, use one of the other methods described above. Write the area below:

_____

_____

5. Enter the data in the appropriate space on the Rocket Tracking and Performance Table on page 198.

Finalizing Your Design

You have learned a great deal about rocket science. Now you can apply what you've learned to modify your preliminary work.

Make any needed changes in your chosen rocket design. Each team member should then prepare full-size drawings of your team's rocket, including orthographic views of at least one of the fins. You can then use this drawing as a template to trace the shape of the fin when you construct it.

THINKING CRITICALLY

1. Observe the fin designs used by your classmates and compare them to your own design. Based on your observations, which rocket do you predict will perform the best? Why?

_____

_____

2. Under what other circumstances do you think estimating skills might be useful when designing rockets?

_____

_____

Modeling

During the modeling part of the project, you will form the body of the rocket, cut out the stabilizing fins and attach them to the body, and select and shape materials for the nose cone. You will then construct various rocket recovery systems according to your initial design ideas and test them to determine which design will probably provide the longest hang time. Once the main part of the rocket has been completed, you will install the engine and the recovery system and test the rocket for stability, making design modifications as necessary.

ROCKET CONSTRUCTION

For this part of the project, you will be constructing the rocket body tube using lamination techniques. Lamination involves bonding layers of materials together with adhesives. Your teacher will distribute samples of laminated materials and products for you to examine.

MATERIALS AND EQUIPMENT

Quantity	Description
1	model rocket engine
1	engine holder tube
1	18 gauge wire or large paper clip, 100 mm long
2	spacing rings for engine holder tube
1	wooden dowel (diameter will depend on diameter of rocket tube)
1 sheet	brown kraft paper
1 roll	tape, 3 in. (75 mm) wide
as needed	hot glue sticks or white glue
1 pair	scissors
assorted	corks
1	plastic soda straw
assorted	materials for fins (balsa wood, styrene sheet, pine, paperboard, etc.)
1	X-acto® knife
assorted	abrasives

Continued

Quantity	Description
1 per 3 teams	fin alignment and assembly fixture (Fig. 6-21 on p. 000.)
assorted	materials for recovery system, such as nylon cloth or Tyvek®
assorted	string, rubber bands, or elastic for shock cord
1	stapler
1	stopwatch
1	flameproof wadding
1 piece	1 string, approximately 152 cm long
1	ruler or straightedge for determining center of mass
assorted	paints, brushes, and other supplies

SAFETY FIRST

You will be using sharp cutting tools and adhesives during this activity. Be sure you understand and follow all safety rules.

You will also need to use fin alignment and assembly fixtures similar to the one illustrated in **Fig. 6-21**. A variety of fixtures must be available in the lab to allow for the varying numbers of fins (usually three or four) that the teams' designs will have. Your teacher may provide the fixtures or may ask you to build your own.

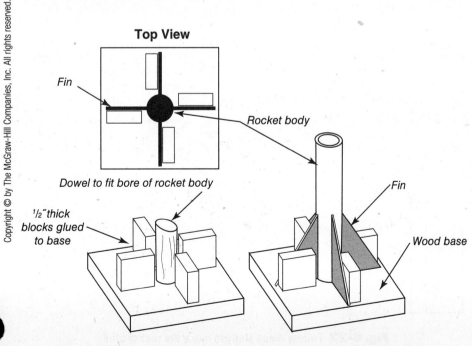

Top View

Fin

Rocket body

Dowel to fit bore of rocket body

Fin

½" thick blocks glued to base

Wood base

Fig. 6-21 Fin alignment and assembly fixture for rockets with four fins

NOTES

Constructing the Rocket Body

Keep in mind that construction quality will play an important part in your rocket's performance. Follow the steps shown in **Fig. 6-22** to construct your rocket tube.

To make sure the rocket will be launched properly, you will need to make a guide, or launch lug. The launch lug slides over a perpendicular rod on the launch pad to hold the rocket straight.

Step 1. Wrap the paper lengthwise around the dowel. The paper will overlap. Glue the overlapping edge of the paper, being careful not to get any glue on the mandrel, which will be removed once the tube has been formed.

Step 2. Cover the paper by wrapping Tape 1 *lengthwise* around the paper-covered mandrel. The edges of the tape will overlap, strengthening the body.

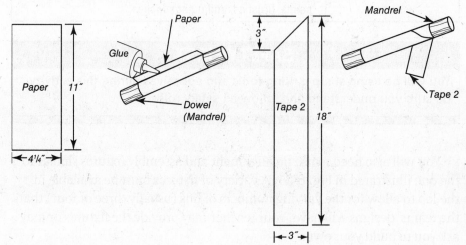

Step 3. Cut one end of Tape 2 at an angle. Place the angle cut at the edge of the paper (which has already been covered by Tape 1) and wrap Tape 2 *diagonally* around the paper body until you reach the end of the paper.

Step 4. After you have reached the end of the paper with the Tape 2 wrapping, slide the rocket body tube you have just formed off the mandrel (dowel) and trim the excess tape.

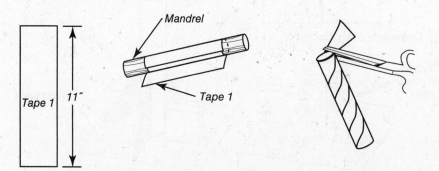

Fig. 6-22 Follow these steps to make the rocket tube.

Project 6 Modeling

1. Cut a 90 mm section from a plastic soda straw.

2. Measure 70 mm *up* from the base of the rocket tube and make a mark. Glue the straw vertically on the rocket body so the bottom of the straw is at this 70 mm mark. Be sure to carefully align the lug on the rocket to insure a straight launch.

Constructing and Attaching Fins

Construct your fins according to your design from the materials you have chosen. Remember that the leading edges of the fins should be rounded over and the trailing edges should be sharp.

Attach the fins to the rocket body tube using hot glue or white glue. It is very important that the fins be accurately aligned on the rocket body. To insure accurate alignment, use a fin alignment and assembly fixture, as shown in Fig. 6-21. After the fins have been attached and the glue has dried, create fillets between the fins and rocket body by smoothing white glue with your fingertips.

Constructing and Installing the Engine Retainer

Figure 6-23 shows how to construct and install the engine retainer if one is not available. The retainer prevents the engine from being pushed up or down the rocket tube during launch.

Step 1. You will need a 100 mm piece of 18-gauge wire or a large, straightened paper clip to make the engine retainer. Cut the engine retainer to length and then form as shown in A.

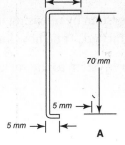

Step 2. Refer to B for this step and the remaining steps. Drill or poke small holes through each side of the rocket body (directly across from each other). Insert the 25mm end of wire through these holes so that the wire passes through the rocket body tube and protrudes beyond this tube.

Step 3. Bend the protruding section of wire that extends beyond the rocket body tube down and firmly press it against the rocket body tube to hold the wire in place.

Step 4. Glue the 70mm length of wire that extends down the rocket body to the rocket body tube. The 5mm end of wire should clip snugly under the base of the rocket body tube.

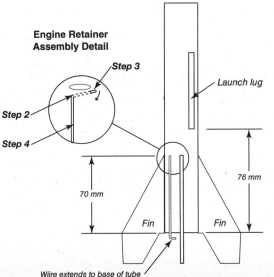

Fig. 6-23 Follow these steps to construct your rocket.

NOTES

Constructing the Nose Cone

Construct and shape the nose cone according to your design. Then glue it to a 12 mm length of wooden dowel that is the same diameter as the dowel used as the mandrel when you made the rocket body tube. The nose cone should then fit easily onto the rocket body.

Making, Testing, and Attaching the Recovery System

Experiment with the different designs you've been considering for the recovery system. Remember that the recovery system must be made of materials that will *not* support combustion; that is, they should not be flammable.

1. Determine which design will probably provide the best solution and do any necessary testing.

2. Make any needed modifications to improve the recovery system.

3. Be sure the recovery system can slide easily into and out of the rocket tube. It should not be too big. The smoothness of the payload bay area within the rocket tube will also affect how easily the recovery system can be deployed. Take measures to make sure this area is as smooth as possible.

4. Install the recovery system in the rocket tube and blow through the tube. If you cannot blow the recovery system out of the tube, it will probably not be deployed by the ejection charge.

5. Test the recovery system by releasing it from a balcony into the air with the rocket attached. Time how long it takes to reach the floor.

6. Attach the recovery system to the rocket and the nose cone, as shown in **Fig. 6-24**.

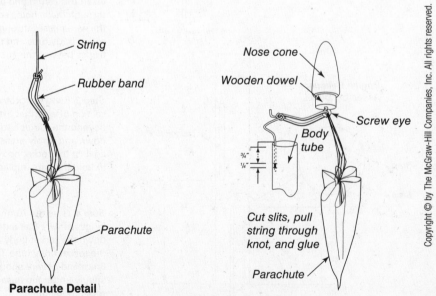

Parachute Detail

Fig. 6-24 Here is an installation for a recovery system. **Yours could be different.**

Testing Rocket Stability

No rocket will be launched if it proves to be unstable, since it would be a safety hazard. The techniques for testing stability are illustrated in **Figs. 6-25** and **6-26**.

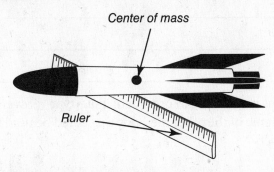

Fig. 6-25 Using a ruler to determine rocket's center of mass.

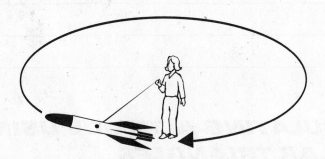

Fig. 6-26 A length of string us taped to the rocket's center of mass, and then the string is used to swing the rocket in a path around one's body to determine the rocket's stability.

1. Determine your rocket's center of mass by balancing it on the edge of a ruler. Mark the center.

2. Securely tape a piece of string to that point on the rocket body.

3. Use the string to hold the rocket as you twirl it around your body. This will determine the rocket's stability.

4. If the rocket is unstable, brainstorm what might be done to make it stable, and make the needed changes. For example, you might increase the area of the fins, streamline the shape of the nose cone, or remount the fins to the body so they are evenly placed. Also, weight could be added to the rocket below the center of mass, but this would reduce performance.

Applying the Finish

Finish your rocket as desired. Remember that a smooth finish, achieved with coats of paint or other material, reduces drag. Thus, the smoother the finish, the less drag there will be. However, keep in mind that the paint or other material will add weight.

Project 6 Modeling

NOTES

THINKING CRITICALLY

1. Would larger fins provide greater stability or less stability in rocket flight? What other effects might result from using larger fins?

_____

_____

_____

2. Some teams may decide to use a parachute-type recovery system. What are some of the other methods that could be used to accomplish recovery?

_____

_____

_____

_____

CALCULATING ALTITUDE USING SIMILAR TRIANGLES

Suppose you have just launched your rocket. The distance from the tracking station to the launch pad is 150 meters, and the tracker spots your rocket at an angle of 52° above the horizontal. How high is the rocket?

MATERIALS AND EQUIPMENT

Quantity	Description
1 per student	ruler
1 per student	protractor
1 per student	calculator

To find the altitude, you will first need to construct a similar triangle. In geometry, "similarity" is a term describing figures that are the same shape but are not necessarily the same size or in the same position.

1. Refer to **Fig. 6-27**. Carefully measure the lengths of the sides of the triangles in the drawing in centimeters. Then set up appropriate ratios and proportions. For example, in Fig. 6-27: $c_1/c_2 = a_1/a_2$, and $c_1/a_1 = c_2/a_2$.

2. There are three sides in each triangle. When two triangles are similar, how many different ratios and proportions could be written?

_____

_____

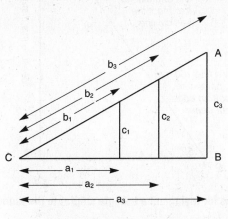

Fig. 6-27 Similar triangles

3. Refer to **Fig. 6-28** showing the launch path of the imaginary rocket described above. Draw the triangle shown in **Fig. 6-29** on page 182 and measure its height.

_____

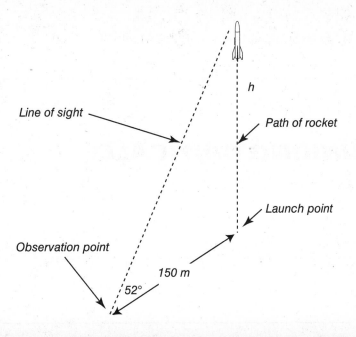

Fig. 6-28 Launch site with the apogee (highest point) of the rocket flight observed at an angle of 52° from horizontal

Project 6 Modeling

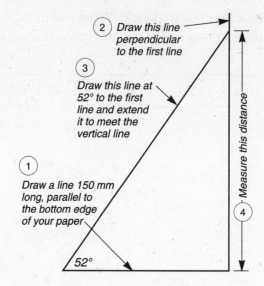

Fig. 6-29 Steps for laying out a triangle similar to the launch site triangle

4. The triangle you have drawn and the launch site triangle are similar triangles. The ratio between the known distance from the observation point to the launch pad (150 m) and the altitude that the rocket reached (h) is the same as the ratio between the 150 mm side of the triangle and the side that is perpendicular to it:

$$\frac{150 \text{ m}}{h} : \frac{150 \text{ mm}}{92 \text{ mm}}$$

What is the altitude the rocket reached?

_____

_____

THINKING CRITICALLY

1. How is the altitude that the rocket reaches related to the horizontal distance the rocket lands from the launch pad? Does the wind speed affect the answer to this question?

_____

_____

_____

_____

2. The wind causes the rocket to assume a course that is not vertical. Instead, it heads at an angle away from the observation point. Will the elevation angle recorded overestimate or underestimate the actual height?

NOTES

_____

_____

_____

CLASSIFYING TRIANGLES

Many angles are involved in tracking a rocket's trajectory and determining the altitude it reaches. **Figure 6-30** shows the three types of triangles—obtuse, right, and acute.

MATERIALS AND EQUIPMENT

Quantity	Description
1 sheet per team	graph paper with 1cm squares, or similar
1 pair	Scissors
1	right-angle "tool" for comparing angles (This could be made from a piece of cardboard.)
1	calculator

1. Using the right-angle tool, compare the size of an acute and obtuse angle to the size of a right angle in the different triangles. As you will see, an acute angle is smaller than a right angle and the obtuse angle is larger.

2. With your teammate, cut from the graph paper squares made from the following number of units: 1, 4, 9, 16, 25, 36, 49, 64, 81, 100, 121, 144, and 169.

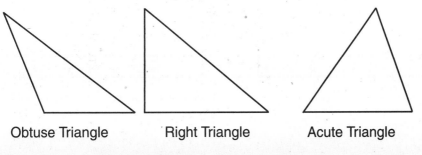

Obtuse Triangle   Right Triangle   Acute Triangle

Fig. 6-30 The three types of triangles

NOTES

3. Put the corners of different-sized squares of graph paper together in such a way that the space left open forms a triangle. (The squares *must not overlap*.) An example is shown in **Fig. 6-31**.

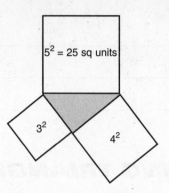

$5^2 = 25$ sq units

3^2

4^2

Fig. 6-31 Constructing a 3-4-5 triangle using paper squares

4. As you form each triangle, fill in the data in the appropriate spaces in the Length and Area Relationships table. The data for the configuration shown in Fig. 6-31 has already been filled in on the top line to show you how it's done. The last column asks you to decide whether the figure formed is a right, acute, or obtuse triangle, or no triangle at all.

LENGTH AND AREA RELATIONSHIPS

Area of largest square	Area of smallest square	Area of middle-sized square	Sum of areas of two smaller squares	Length of longest side of triangle	Length of shortest side of triangle	Length of middle-sized side of triangle	Decision: right, obtuse, acute, or none
25	9	16	25	5	3	4	right

Project 6 Modeling

5. Review the data on your table. What is the relationship between the sum of the areas of the two smaller squares and the largest one when a right triangle is formed?

_____

_____

6. What is the relationship between the sum of the areas of the two smaller squares and the area of the largest one when an acute triangle is formed?

_____

_____

7. What is the relationship between the sum of the areas of the two smaller squares and the area of the largest square when an obtuse triangle is formed?

_____

_____

8. What is the relationship between the sum of the areas of the two smaller squares and the area of the largest one when no triangle is formed?

_____

THINKING CRITICALLY

1. If three sides of a triangle are known, how can you determine if the triangle is a right triangle?

_____

2. If a triangle is a right triangle, could either of the other angles be obtuse?

_____

_____

_____

3. If you know the length of the two legs of a right triangle, how may you find the length of the hypotenuse?

_____

_____

Project 6 Modeling

NOTES
USING SIMILAR RIGHT TRIANGLES

Math

There are also two acute angles in a right triangle. One leg is opposite the angle, and the other leg is adjacent to it.

MATERIALS AND EQUIPMENT

Quantity	Description
several sizes	30°-60°-90° triangles, such as those used for drafting
several sizes	45°-45°-90° triangles, such as those used for drafting
1 per student	ruler
1 per student	protractor
1 sheet per student	graph paper
1 per team	calculator with trigonometric functions (or trigonometric tables)

1. Measure the lengths of the sides of the 30°-60°-90° and the 45°-45°-90° drafting triangles. Identify ratios that are apparent with respect to the sides of each triangle.

30°-60°-90°		45°-45°-90°	
a/c	.5	a/c	.71
b/c	.87	b/c	.71
a/b	.58	a/b	1
c/a	2	c/a	1.41
c/b	1.15	c/b	1.41
b/a	1.73	b/a	1

2. Compare the lengths of the sides of two or more 30°-60°-90° triangles (or the 45°-45°-90° triangles) of different sizes. What conclusions can you draw from the comparisons?

_____

_____

_____

_____

Project 6 Modeling

3. Look at acute angle *A* in **Fig. 6-32.** In the smallest triangle (a_1-b_1-c_1), which is the opposite leg and which is the adjacent leg?

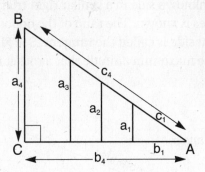

Fig. 6-32 Similar right triangles

_____

4. In this same triangle, what is the ratio of the opposite leg from angle *A* to the adjacent leg?

_____

5. In the largest triangle, what is the ratio of the opposite leg from angle *A* to the adjacent leg?

_____

6. If any other right triangle was drawn that contained angle *A* as one of its acute angles, what would be the ratio of the length of the leg opposite angle *A* to the length of the adjacent leg?

_____

_____

_____

7. If angle *A* were increased in size, would this change the ratios that have been established? If so, how?

_____

_____

_____

8. Are these relations "functions" of the size of the acute angle?

_____

NOTES

Tangents

Knowing the ratio of two sides in a right triangle can be used to solve for a corresponding unknown side in a similar right triangle when one of the corresponding sides is known. The ratio of the opposite side of a right triangle to the adjacent side is called the **tangent**. See **Fig. 6-33**. This ratio can help determine the maximum altitude that a rocket reaches.

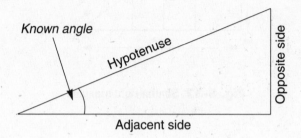

Tangent = ratio of the opposite side to the adjacent side = $\dfrac{\text{Opposite side}}{\text{Adjacent side}}$

Fig. 6-33 Determining the tangent

1. Suppose you have two similar right triangles. The length of the adjacent side to angle B is 10 in one of the triangles. The ratio of the length of the opposite side to the adjacent side of angle B in another triangle is 0.64. How could this information be used to find the length of the opposite side in the first triangle?

_____

You can determine the tangent of an angle using your calculator or a table of trigonometric functions. Look again at Fig. 6-28 on p. **000**. The elevation angle, the point at which the rocket is observed, is 52° above horizontal. Relative to the elevation angle, which is the hypotenuse, the adjacent side, and the opposite side? Mark each of these on the illustration.

3. Identify what is known about this triangle.

_____

_____

The tangent of the 52° angle is 1.28:

$$\text{Tangent} = \frac{\text{Length of opposite side}}{\text{Length of adjacent side}} = 1.28$$

$$\text{Tangent} = \frac{\text{Length of opposite side}}{150 \text{ meters}} = 1.28$$

The height of the rocket can then be determined using this sequence:

Length of opposite side = 150 × 1.28 = 192; therefore,

the height of the rocket = 192 meters.

THINKING CRITICALLY

1. If the elevation angle of a rocket was 45° above horizontal, how could you determine its altitude without using the tangent function?

   _____

   _____

   _____

   _____

2. Is it possible for the altitude of a rocket to be greater than the length of an observer's line of sight from the observer's eye to the rocket?

   _____

   _____

   _____

   _____

NOTES

ROCKET TRACKING SYSTEMS

Your teacher will determine whether the rocket performance data is to be analyzed using similar right triangles or trigonometric functions. If trigonometric functions are to be used, construct a clinometer, as shown in **Fig. 6-34**.

MATERIALS AND EQUIPMENT

Quantity	Description
1	protractor or clipboard
1	soda straw
1 piece	string long enough to suspend the weight below the protractor or clipboard when attached to the top of the protractor or clipboard (Figs. 6-34 and 6-35)
1	weight (such as washer or hex machine nut)

A **clinometer** is an instrument used to measure angles of slope or inclination. In this case, a protractor is used to determine the angle of a launched rocket relative to the horizontal plane of the earth.

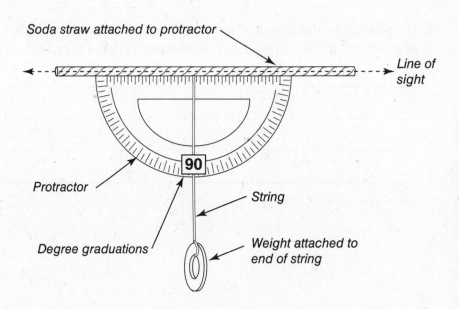

Fig. 6-34 Apparatus for determining the mechanical advantage of the syringe hydraulic system

Project 6 Modeling

If similar right triangles are to be used to analyze rocket performance data, construct a hypsometer-type altitude tracking device. See **Fig. 6-35**. This **hypsometer** is used to mark off angles similar to those of the launched rocket in relation to the earth. Similar triangles rather than direct readings of the graduations on a protractor are then used to determine the height of the launched rocket.

Either of these simple devices will provide rough estimates of the rocket's altitude.

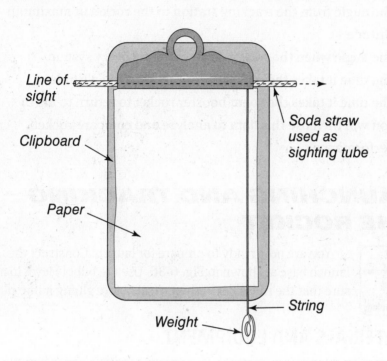

Fig. 6-35 Hypsometer-type altitude tracking device; used when analysis is performed using similar right triangles.

THINKING CRITICALLY

1. What is the purpose of the weight attached to the end of the string on either device?

_____

_____

2. How could the clinometer or hypsometer be made easier to use during the tracking process?

_____

_____

Evaluation

Your rocket will be tested and analyzed during this part of the project. The launch pad will be set up in an open area and the rockets launched. You will collect the required data:

- The angle from the tracking station to the rocket at maximum altitude
- The angle when the rocket deploys the recovery system
- The time it takes the rocket to reach maximum altitude
- The time it takes the spent booster rocket to return to Earth
  You will then use this data to analyze and compare rocket performance.

LAUNCHING AND TRACKING THE ROCKET

Engineering Technology

You are now ready to prepare for launch. Construct the launch base as shown in **Fig. 6-36**. Use the bubble level to make sure that the rod is perfectly vertical before gluing it into place.

MATERIALS AND EQUIPMENT

Quantity	Description
1	launch base of ¾″ plywood, 25 cm × 25 cm (or larger)
1	drill with small-diameter bit
1	welding rod, about 1 meter long and of a diameter that will fit loosely inside the soda straw launch lug
1	bubble level
1 tube	adhesive
1	completed rocket with engine installed
1	ignition switch for launcher with safety key, wire, and clips
1	scale
1 per rocket	rocket igniter
1	tape measure (100 feet or 30 meters long)
2	clinometers, hypsometers, or surveying transits
2	stopwatches

*1 m welding rod—
vertical with base*

*Rocket with
engine installed,
ready for launch*

*Soda straw launch
lug on launch rod*

25 cm x 25 cm base

*Drill hole slightly smaller than rod—push rod in for tight fit.
Glue in place*

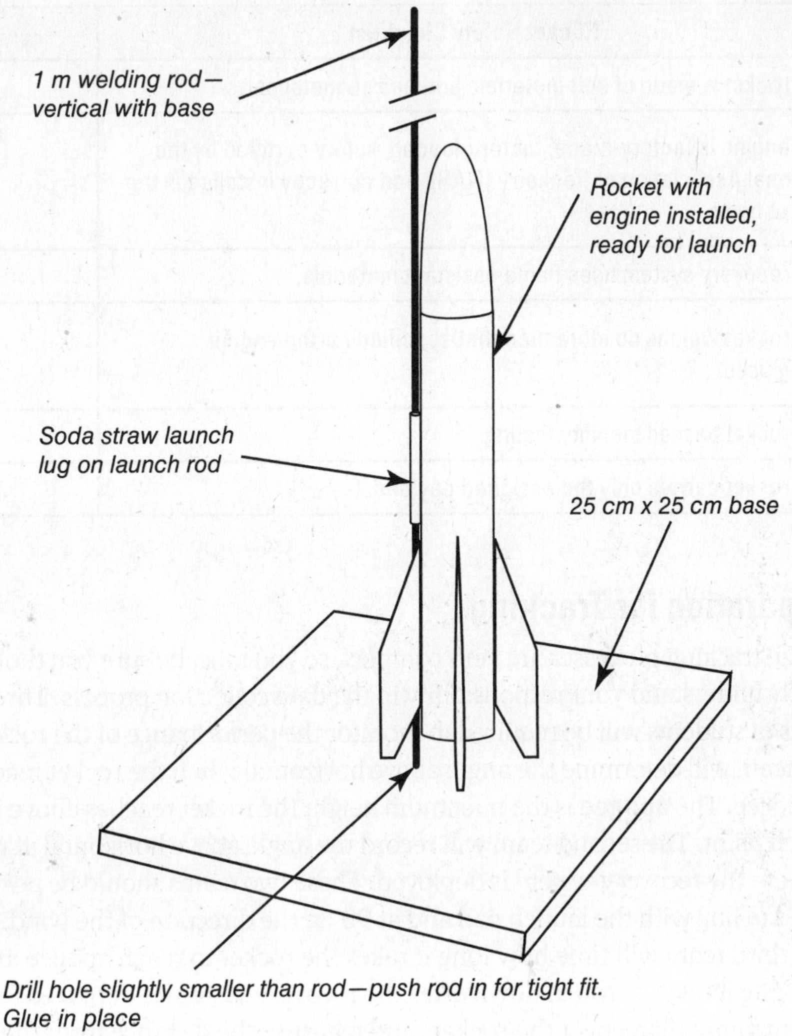

Fig. 6-36 Setting up the launch base

Go over the Rocket Safety Checklist on page 194. One point that must be checked relates to the weight of the rocket. Weigh your rocket and enter its weight on the checklist and on the Rocket Tracking and Performance Table on page 198.

SAFETY FIRST

Safety Checklist After weighing your rocket, show the rocket, the Rocket Tracking and Performance Table, and the Rocket Safety Checklist to your teacher for approval.

Project 6 Evaluation

NOTES

Rocket Safety Checklist	√
The rocket is made of soft materials and has no metal parts.	
The engine is factory-made, factory-loaded, safety certified by the National Association of Rocketry (NAR), and correctly installed in the rocket body.	
The recovery system uses flame-resistant materials.	
The rocket weighs no more than that specified for the engine being used.	
The rocket passed stability testing.	
The rocket carries only the assigned payload.	

Preparation for Tracking

The tracking processes are very complex, so you must be sure you thoroughly understand your responsibility in the data collection process. Three teams of students will be required to monitor the performance of the rockets. One team will determine the angle above horizontal when the rocket reaches its apogee. The **apogee** is the maximum height the rocket reaches above its launch point. The second team will record the angle above horizontal at the moment the recovery system is deployed. These two teams should be positioned in line with the launch pad and at 90° to the direction of the wind.

A third team will time how long it takes the rocket to reach apogee and how long it takes to touch down after the recovery system is deployed. Other teams will recover the rockets and measure the distance from the launch point to the touchdown point. Your teacher will act as the launch director and will ignite the rocket engines.

Tracking teams should be located at least 60 m (about 200 ft.) from the launch pad. This is known as the *baseline distance*.

Using the Clinometer or Hypsometer

When reading the clinometer or hypsometer, the weighted string should be pressed against the protractor or clipboard at the appropriate times during the flight—at apogee and at deployment of the recovery system. For example, when the rocket is at apogee, press the string and read the marking the string crosses. Suppose the clinometer reads 70°. This means that the angle at apogee equals 90° minus 70° (20°). The hypsometer captures the angle in a similar manner, and it is marked on the paper attached to the clipboard. The difference is that the clinometer uses direct readings of the graduations on a protractor, and the hypsometer uses similar triangles.

Figure 6-37 shows how a clinometer (or hypsometer or surveying transit) is used to track each rocket and make accurate measurements. Record your data on the Rocket Tracking and Performance Table on page 198.

Name________________________ Date______________ Class ______________

NOTES

ROKV11V2

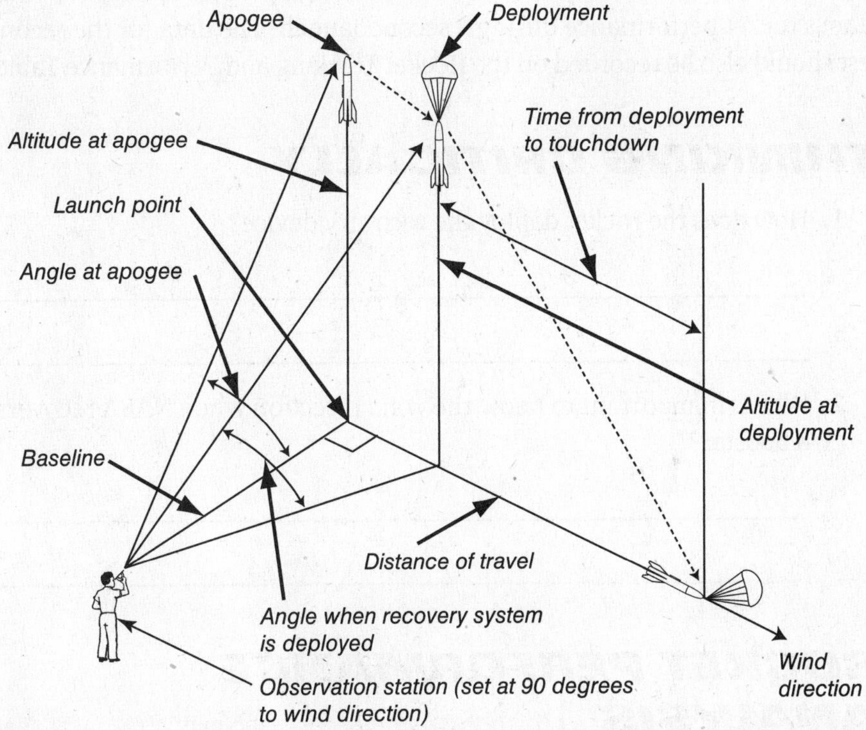

Fig. 6-37 Tracking the rocket's flight path

Countdown and Launch

The tracking and recovery crews should take their assigned positions and caution any other observers to stay outside the launching circle. Prepare to launch your rocket by following the steps below.

1. Pack the recovery device with flameproof wadding.
2. Install the payload and nose cone.
3. Check the engine holder and relative placement of multistage engines.
4. Secure the igniter.
5. Secure the engine.
6. Install the rocket on the launcher and connect clean microclips.
7. Clear the area and alert the launch, tracking, and recovery crews. Check for any low-flying aircraft.
8. Arm the launcher by inserting the safety pin/key.
9. Sound the ignition warning.
10. Ask your teacher to ignite the rocket.
11. Count down to ignition.

Modification and Retesting

After the rockets have been launched and recovered and the performance data has been recorded, begin the redesign process to improve your team's rocket performance during a second launch. The data for the second test should also be recorded on the Rocket Tracking and Performance Table.

THINKING CRITICALLY

1. How does the rocket deploy the recovery device?

   _____

   _____

2. Why is it important to know the wind direction when NASA recovers a booster?

   _____

   _____

ROCKET PERFORMANCE ANALYSIS

If you used the hypsometer, you will have the angles drawn on the paper used during tracking. Using this information, determine the altitude at apogee and the altitude at which the recovery system was deployed. Your knowledge of similar triangles will aid you. Record this data on the Rocket Tracking and Performance Table on page 198.

MATERIALS AND EQUIPMENT

Quantity	Description
1 per student	calculator
1 per student	Rocket Tracking and Performance Table (including angles recorded in the field)
1 per team	hypsometer page (if used) showing angle at apogee and angle at deployment of recovery system
1 per team	table of trigonometric tangents (if a clinometer was used)

If a clinometer-type tracking system was used, determine the altitude at apogee and the altitude at which the recovery system was deployed by using a trigonometric table of tangents. Record this data on the Rocket Tracking and Performance Table.

Calculating Velocity

Calculate the average velocity of your rocket from launch to apogee using this formula:

[average velocity] $= \frac{d}{t}$, where t is the measured and recorded time and d is the altitude at apogee or altitude at deployment, respectively.

Record the results on your Rocket Tracking and Performance Table. Then calculate the average velocity from the moment the recovery system was deployed until touchdown using the same formula. Record this data on the Rocket Tracking and Performance Table.

Determining Performance Factors

As a class, prepare a Rocket Performance Factor Chart. See **Fig. 6-38** for an example. Each team should provide the information needed—altitude at apogee and hang time (time from deployment of the recovery system to touchdown)—for their rocket. Then determine the "performance factor" for each rocket by multiplying the maximum altitude reached (in meters) by the hang time (in seconds). Record this data on the chart so you can compare the performances of all the rockets. The rocket with the highest performance factor is the winner of The Challenge.

Rocket Performance Factor Chart			
Rocket Number or Team Name	Altitude at Apogee in Meters	Hang Time in Seconds	Performance Factor
Rocket # 1	170	22	3740
Rocket # 2	170	28	4760
Rocket # 3	200	28	5600

Fig. 6-38 This is a sample Rocket Performance Factor Chart.

Rocket Performance Factor Chart			
Rocket Number or Team Name	Altitude at Apogee in Meters	Hang Time in Seconds	Performance Factor

Project 6 Evaluation

NOTES

THINKING CRITICALLY

1. What factors do you think may account for the differences in performance among the rockets?

_____

_____

2. Name some ways in which mathematics is important in the aerospace industry.

_____

_____

ROCKET TRACKING AND PERFORMANCE TABLE

	Launch 1	Launch 2
Pre-launch Data		
Weight in grams		
Number of fins		
Area of each fin in cm^2		
Distance from observation point to pad		
Launch Data		
Angle at apogee		
Angle at recovery system deployment		
Time from launch to apogee		
Time from deployment to touchdown		
Mathematical Analysis		
Height at apogee in meters		
Height at recovery in meters		
Average speed to apogee		
Average speed from deployment to ground		

Project 6 Evaluation

Pollution-Free Vehicle

Design Brief

In cities all over the United States, air pollution from automobile emissions is a serious problem. Asthma and other breathing difficulties are related to air pollution. In addition, global climate changes have been linked to air pollution. To help solve these problems, automobile manufacturers are encouraged to develop non- or low-polluting vehicles. See **Fig. 7-1**. They are always looking for new ideas.

Challenge

Design and construct a pollution-free vehicle (PFV) that will travel a specified distance and then come to a complete stop at a designated spot.

Fig 7-1 This car, made by MDI of Spain is non-polluting. It runs on compressed air!

Criteria

- Your vehicle must be powered by a self-contained mechanical system that will convert potential energy into kinetic energy without polluting the air.
- During testing, it must travel 8 meters (approximately 25 feet) in a straight line and come to a complete stop.
- It must travel the distance in as short a time as possible.
- All designs must be accompanied by appropriate documentation, including:
  - ☐ Sketches of all possible solutions considered
  - ☐ A final drawing of your chosen solution
  - ☐ Data collected from various tests and experiments
  - ☐ A chart/graph showing how your solution performed
  - ☐ Information gathered from resources
  - ☐ Notes made along the way

NOTES

Constraints

- Your vehicle cannot use CO_2 cartridges or store-bought batteries.
- Your vehicle can be no more than 60 cm in length (about 24 inches), no more than 20 cm in width (about. 8 inches), and no more than 60 cm in height (about 24 inches).

Engineering Design Process

1. **Define the problem.** Write a statement that describes the problem you are going to solve. The Design Brief and Challenge provide information.
2. **Brainstorm, research, and generate ideas.** The Research & Design section, beginning on page 201, provides background information and activities that will help with your research.
3. **Identify criteria and specify constraints.** Refer to the Criteria and Constraints listed on pages 199 and 200. Add any others that your teacher recommends.
4. **Develop and propose designs and choose among alternative solutions.** Remember to save all the proposed designs so that you can turn them in later.
5. **Implement the proposed solution.** Once you have chosen a solution, determine the processes you will use to make your pollution-free vehicle. Gather the tools and materials you will need. Make sure you understand and follow all safety rules.
6. **Make a model or prototype.** The Modeling section, beginning on page 232, provides instructions for building the vehicle.
7. **Evaluate the solution and its consequences.** The Evaluation section, beginning on page 236, describes how to test the vehicle.
8. **Refine the design.** If instructed by your teacher, make changes to improve your design.
9. **Create the final design.** If instructed by your teacher, make a new vehicle based on the revised design.
10. **Communicate the processes and results.** Write a report about your project. Be sure to turn in to your teacher all the documentation listed under Criteria.

Research & Design

In this section you will learn about fundamental technology, science, and mathematics concepts and techniques that relate directly to the design of your pollution-free vehicle. You will draw upon these ideas as you develop your concept of a vehicle and power train.

POTENTIAL AND KINETIC ENERGY

Science

During this activity you will learn how energy is stored and how potential energy is converted to kinetic energy. You will also be able to distinguish among the six forms of energy.

MATERIALS AND EQUIPMENT

Quantity	Description
assorted	springs, rubber bands, and other sources of energy that might be used to power the vehicle
1	wind-up clock

Energy is the capacity to do work. Energy that is stored is called **potential energy**. A hammer left lying on a table, for example, has potential energy. When that hammer is used to pound a nail, its energy is released. See **Fig. 7-2**. Energy released, or in motion, is **kinetic energy**.

Many objects have potential energy. The head of a match contains potential energy. What happens when that match is struck and its energy is released? Can you think of other examples?

Fig. 7-2 As it pounds a nail, the hammer's potential energy is released.

Name_____ Date_____ Class _____

NOTES

All energy in nature can be grouped into six forms:
- Mechanical energy is the energy of motion.
- Electrical energy is the flow of free electrons from atom to atom.
- Nuclear energy is contained within the nuclei of atoms.
- Thermal energy is heat energy.
- Chemical energy is found in fuels and explosives.
- Light energy radiates outward in all directions.

Energy cannot be created or destroyed, but it can be changed from one form to another. When you strike a match, its chemical energy is changed into thermal and light energy.

When energy is converted from one form to another, some is lost into the environment as heat. The heat you feel on the hood of a car after the engine has been running is the result of an energy conversion.

1. Experiment with springs, rubber bands, and other sources of energy that might be used to power your pollution-free vehicle.

2. Note any sources of energy you might use for your vehicle design.

3. Some sources release large amounts of energy at the beginning of their use and small amounts at the end. For example, a rubber band has a large amount of potential energy when it is stretched taut. Then, as it is released, that energy quickly disappears. Brainstorm how an elastic energy source might be controlled so its energy is converted from potential to kinetic at a steadier rate. Study the wind-up clock. Why does it not run faster after the spring has been wound?

THINKING CRITICALLY

1. Why is it important to know about potential and kinetic energy in designing the pollution-free vehicle?

_____

_____

_____

_____

2. Of the six forms of energy that you learned about, what is probably the best one to use in the pollution-free vehicle? Why?

_____

_____

_____

_____

POWER SOURCES

NOTES

For this activity you will discover how some power sources are pollution free and how others pollute the environment. You will learn how different power technologies operate and convert potential to kinetic energy and can be adapted for a vehicle.

Your teacher will divide the class into groups and demonstrate how certain power sources work.

MATERIALS AND EQUIPMENT

Quantity	Description
1	examples of power sources
1	multimeter
2	electric toy motor with low voltage and amperage
2	electric fans

1. An internal combustion engine converts potential chemical energy (gasoline) into thermal energy and finally mechanical energy. Note the power output shaft of the engine, where potential energy becomes kinetic energy. Observe the air pollution that is caused by the engine.

2. As you observe the non-polluting power devices, watch for evidence of potential energy changing into kinetic energy. Identify the mechanisms that store the potential energy.

3. Working with your group, consider each of the examples that was demonstrated to the class. Operate the devices in the same way they were demonstrated. Identify the sources of potential energy for each. List the device and its source of potential energy in the Potential Energy Sources table. Use a separate sheet of paper if necessary.

POTENTIAL ENERGY SOURCES

Device	Potential Energy Source	Polluting or Not?
Wind-up toy motor	Spring	Not

4. Consider the size and weight of each device.

5. Identify those that pollute the environment and those that do not. Determine if any of these non-polluting power sources would possibly make suitable power sources for your pollution-free vehicle.

N O T E S   *THINKING CRITICALLY*

1. Name two things that could cause pollution in a nuclear power plant.

   _____

   _____

2. How could you tell that the internal combustion engine was converting potential energy into kinetic energy?

   _____

   _____

3. How can you tell when gasoline is being converted into thermal energy when an automobile operates?

   _____

   _____

4. How is a wind-up clock powered?

   _____

   _____

5. What powers an electric motor connected to a solar cell?

   _____

6. How can you tell if a battery is producing electricity?

   _____

   _____

7. What power source will most likely be used for a pollution-free automobile?

   _____

   _____

   _____

8. Are vehicles powered by only electric storage batteries really "non-polluting"? Why or why not?

   _____

   _____

   _____

   _____

SIMPLE MACHINES

Science

During this activity, you and your teammates will learn how simple machines are used to change direct, force, and motion. They can make it easier to do work by providing **mechanical advantage**. The six simple machines include the lever, the wheel and axle, the pulley, the inclined plane, the wedge, and the screw. (You can see examples in Fig. 5-2 on page 117.) These machines make work easier by multiplying human force.

MATERIALS AND EQUIPMENT

Quantity	Description
1	letter opener
1	lever
1	wedge
1	scrap of wood into which a nail has been driven
1	claw hammer
2	screwdrivers (one small-diameter handle, one larger)
1	scrap of softwood into which screws can be driven
assorted	wood screws
1	wheel and axle
1	meter stick with fulcrum and weight attachments
1	weight to which a spring scale can be connected, approx. 1 kg
1	inclined plane
1	spring scale
1	fulcrum
1	pulleys of equal diameters
1	pulley with a diameter larger than the other pulleys

Work is using a force to act on an object in order to move that object in the same direction as the force. Knowledge of simple machines allows people to design mechanical devices that require less effort to accomplish work. For example, by using a lever, we can move a heavy object with a force that is less than the weight of the object. See **Fig. 7-3**. The **effort force** is the force a person applies to a machine. The force offered by an object that the machine is acting upon is called the "resistance force." If a person is trying to move a rock with a lever, the rock is the **resistance force**, and the person trying to move the rock is applying the effort force.

NOTES

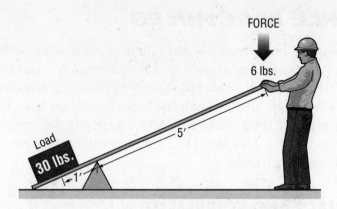

Fig. 7-3 A lever provides mechanical advantage.

You can find the work done to lift an object against the force of gravity by multiplying its weight (resistance force) by the distance it was lifted. If you apply an effort force but fail to move an object, you may be tired from the effort, but you did not do any work.

1. Your teacher will distribute devices based on simple machines to groups of students. Working with your group, use the device to discover how it provides mechanical advantage. Exchange devices with other groups so you try each one. Here are some examples:
   - Try to pull a nail out of a piece of wood with your fingers. Then use the claw of a hammer.
   - Try to drive a screw into a piece of soft wood with your fingers. Then use a screwdriver with a small-diameter handle. Next use a screwdriver that has a larger-diameter handle.

The Inclined Plane

2. A spring scale is attached to a 1 kg weight. As you work with the inclined plane, fill out the Inclined Plane Advantage table.

INCLINED PLANE ADVANTAGE

Lifting Weight Vertically			Pulling Weight along Inclined Plane		
Distance in cm	Force	Force x distance =	Length of inclined plane in cm	Force	Force x distance =

3. Holding the spring scale, one student should attempt to lift the weight straight up from the floor to the top of the inclined plane in five seconds or less. Another student should note the reading on the spring scale as the weight is lifted. Enter the force and distance from the floor to top of the inclined plane in the Inclined Plane Table.

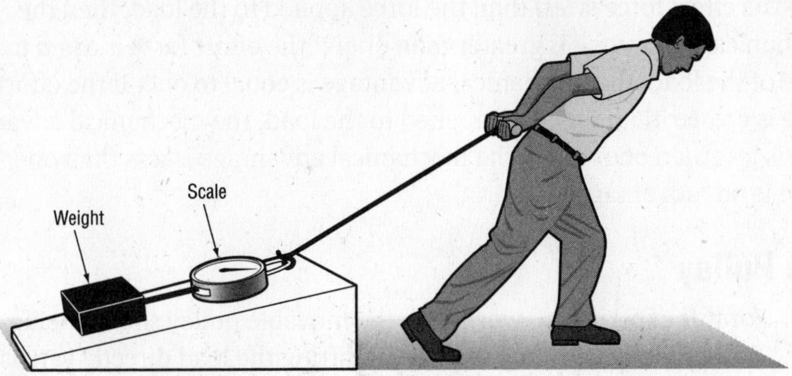

Fig. 7-4 Weight and scale being drawn up an inclined plane

4. Holding the spring scale, the first student should drag the same 1 kg weight up the inclined plane in five seconds or less. See **Fig. 7-4**. The second student should note the reading on the spring scale as before. Enter these data in the table.

Since the 1 kg weight was lifted through the same vertical distance, the same amount of work was done each time. However, the effort force was less when pulling the weight up the inclined plane than it was lifting the weight straight up.

5. Determine the amount of work done by multiplying the weight of the object by the distance through which it was moved. Enter these values in the table. Are the two calculations equal? Why or why not?

_____

_____

6. Connect the spring scale to one end of a meter stick, and connect a 1 kg weight to the other end. Place the middle of the meter stick on a fulcrum. What is the mechanical advantage?

_____

_____

7. How does the distance through which the force is applied compare to the distance that the load moves?

_____

8. Experiment with placing the fulcrum at different places along the meter stick. Calculate the mechanical advantage in each case.

NOTES

If the effort force is *less* than the force applied to the load, then the mechanical advantage is greater than one. If the effort force is *equal* to the force of the load, the mechanical advantage is equal to one. If the effort force is *greater* than the force applied to the load, the mechanical advantage is less than one. When the mechanical advantage is less than one, there is no "advantage."

The Pulley

9. For this experiment, work with the movable pulley shown in **Fig. 7-5**. Try to predict the force required in lifting the load directly versus using the pulley.

_____

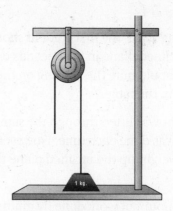

Fig. 7-5 Fixed pulley

As you use the pulley to lift the load a distance of 25 cm, note the direction in which the effort force is applied and the direction in which the load moves.

_____

11. Measure the forces required with the spring scale. Also measure the distance through which the effort force is applied and the distance through which the resistance force moves. Use this information to fill out the Movable Pulley Results table.

FIXED PULLEY RESULTS

Effort Distance in cm	Force	Resistance Distance in cm	Force x distance =

12. Given the distances and forces, determine the amount of work done. Write that amount in the table.

13. Install a larger pulley in the pulley apparatus. Predict whether the large or small pulley requires less work.

NOTES

_____

_____

Now try lifting the same weight. Was your prediction correct?

_____

The effort force is equal to the resistance force regardless of the size of the pulley, so no force advantage is gained. However, the effort force and the resistance forces move in opposite directions. Can you think of ways a fixed pulley might be used in the vehicle you will be designing? Name some practical examples in which it is necessary to use a pulley in order to have the effort force move in the opposite direction to the resistance force.

_____

_____

14. Work with a movable pulley as shown in **Fig. 7-6**. Compare lifting the load a distance of 25 cm with the pulley compared to lifting the load directly as you did in Steps 9 and 10. Then use a spring scale to measure the force applied to raise the load through the 25 cm distance. Use the data you collect to fill out the Movable Pulley Results table.

Fig. 7-6 Movable pulley

Name_____ Date_____ Class_____

MOVABLE PULLEY RESULTS

Effort Distance in cm	Force	Resistance Distance in cm	Force x distance =

How did using the movable pulley compare to using the fixed pulley?

_____

_____

15. Add a fixed pulley of equal size to the fixed pulley arrangement. See **Fig. 7-7**. Predict which of the three pulley arrangements will require the least amount of work.

_____

_____

_____

_____

_____

_____

_____

_____

Use the pulleys to raise the load to the top. Record your data on the Double Fixed Pulleys Results table. Assess whether your prediction was correct.

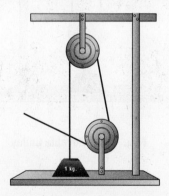

Fig. 7-7 Two fixed pulleys of equal size

DOUBLE FIXED PULLEYS RESULTS

Effort Distance in cm	Force	Resistance Distance in cm	Force x distance =

There is a tradeoff in using machines that have a mechanical advantage. Though such machines allow you to move a load with a force less than that of the load, the effort force must always be applied through a greater distance than the load will actually move.

THINKING CRITICALLY

1. Is there any advantage in using a machine that has a mechanical advantage of less than one?

_____

_____

_____

_____

2. Dragsters are very powerful, specially designed racing cars. One of the problems with them is that the front end of the dragster tends to lift up off the road when power is applied. Based on your knowledge of mechanical advantage, how would you design a dragster to limit this lifting action during a race?

_____

_____

The chassis of the dragster is a lever arm and the rear axle is the fulcrum. One solution is to lengthen the chassis of the dragster and move the engine forward. More force would therefore be required to lift the front of the dragster.

NOTES

TRANSMISSION AND CONTROL MECHANISMS

Engineering Technology

For this activity you will experiment with devices that transfer motion, change the direction of motion, and create mechanical advantage. You will find ways to adapt the mechanisms for use in your pollution-free vehicle.

MATERIALS AND EQUIPMENT

Quantity	Description
1	mousetrap
2 feet	string
1 set	wheels mounted on an axle

SAFETY FIRST

Before You Begin Make sure you understand how to handle equipment and materials safely. Ask your teacher to demonstrate their proper use. Unplug any power equipment while it is being examined. Follow all safety rules. Refer to the Safety Handbook, beginning on page 291, for more information about safety in the lab.

1. Your teacher will introduce you to several devices, including equipment used in the lab. Try to identify the simple machine each example uses to transmit power. As you observe, fill in the Simple Machines That Transmit Power table. The first entry is done for you.

SIMPLE MACHINES THAT TRANSMIT POWER

Lab Device	Simple Machine
screwdriver	wheel and axle

Many of the power sources you have studied have **transmissions**— mechanisms that transmit power and motion. For example, the wind-up toy motor has gears inside that transmit the power of the spring. Gears are similar to pulleys but do not require a belt to transmit energy. If the motor is connected to a set of wheels, then it will turn the wheels. All of the possible power sources for your pollution-free vehicle are useless unless the kinetic energy they produce can be transmitted to the vehicle wheels.

For example, the different sets of pulleys inside the band saw transmit the motion and power of the motor to the saw blade. The small pulley mounted on the motor turns quickly, but the large wheels over which the blade passes turn more slowly. Large wheels, pulleys, and levers can provide more **torque** (rotational force) than smaller ones because they are better multipliers of force.

2. Set up a mousetrap with the wheel-and-axle assembly shown in **Fig. 7-8**.

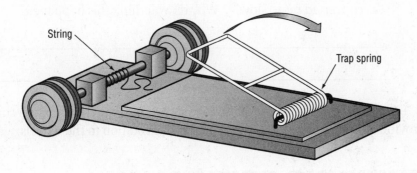

String

Trap spring

Fig. 7-8 Mouse trap and wheel assembly

3. Retract the spring on the mousetrap and release it. Note that the potential energy of the spring was used up very quickly.

4. This time, tie the string to the lever of the mousetrap and to the axle of the wheel assembly. See **Fig. 7-9**. Wind the slack portion of the string around the axle until the string is tight. Set and release the trap.

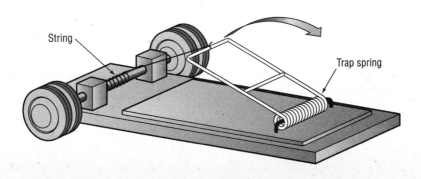

String

Trap spring

Fig. 7-9 Using the set-up as a transmission and load.

Name_____ Date_____ Class _____

5. Which part of this assembly acts as a transmission?

_____

6. Linkages are connecting mechanisms that often apply levers. Which part of the mousetrap device provides a linkage?

_____

7. Which part of the assembly acts as a pulley?

_____

8. Would an axle with a larger diameter provide more or less force to the wheels?

_____

9. Did you notice that this time the energy from the mousetrap spring was converted more slowly? Why do you think this happened?

_____

_____

Depending on which way the string was wound, the transmission could provide either clockwise or counter-clockwise rotation to the wheels.

THINKING CRITICALLY

1. How can the energy in a spring be controlled so that it is not released too quickly?

_____

_____

_____

_____

2. What method could you use to increase the energy of the power unit?

_____

_____

DETERMINING ELASTIC POTENTIAL ENERGY

During this activity, you will learn how springs and rubber bands store and release energy. You will use what you learn to adjust the energy systems in your vehicle.

MATERIALS AND EQUIPMENT

Quantity	Description
5	small weights
1	rubber band or spring
1	rule
1	safety goggles

SAFETY FIRST

Before You Begin Springs and rubber bands stretched too tightly can break. Be sure to wear safety goggles during this experiment.

The potential energy stored in a spring or rubber band depends on the length to which it is stretched. The **elongation** is the difference between the original length of the rubber band or the spring and its length after a given force, or load, is applied. The **spring constant** *(k)* tells how "stiff" it is. The spring constant is determined by dividing the applied force *(F)* by the elongation:

$$k = F/x$$

1. Measure the length of a spring or rubber band. If a rubber band is used, straighten it out with your finger tips before measuring it. Record the length:

2. Hold one end of the spring or rubber band in a fixed position by looping it over a pencil or similar device. Add weights in even increments to the other end, recording the weight applied and the length of the spring or rubber band on the Determining the Spring Constant table. Repeat this six times, increasing the weight until the spring or rubber band approaches its breaking point (elastic limit). See **Fig. 7-10** on page 216.

3. To determine the spring constant, divide the force, or weight, by the elongation produced by that force: $k = F/x$. Do this calculation for each trial and take the average in order to obtain an approximate spring constant value.

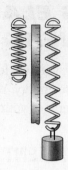

Fig. 7-10 Calculating the spring constant

DETERMINING THE SPRING CONSTANT

Trial	Force (weight) F	Elongation x	Spring Constant K = F/x
1			
2			
3			
4			
5			
6			
Average			

4. The total potential energy (U) of the spring or rubber band is equal to half the product of the spring constant (k) times the elongation (x) squared, as follows:

$$U = 1/2 \ kx^2$$

Using the formula to determine the potential energy, we would have:

$U = 1/2 \ kx^2$ $U = 1/2 \ (240)$

$U = 1/2 \ (3.75)(8^2)$ $U = 120$

$U = 1/2 \ (3.75)(64)$

THINKING CRITICALLY

1. How can you increase energy stored by a spring or rubber band?

_____

_____

2. Would you expect the spring constant to increase or decrease with wear on the spring?

_____

_____

_____

CIRCUMFERENCE

During this activity you will determine the circumference of a wheel when the diameter or radius is given and determine the diameter or radius when the circumference is given. This will be useful when calculating the distance your vehicle will travel.

MATERIALS AND EQUIPMENT

Quantity	Description
1 set	wheels of various sizes, 10 cm minimum
1	calculator
1 large sheet	graph paper
1	metric tape measure
1	dividers

1. Place a single mark on the edge of one of your wheels. Turn the wheel so that the mark is next to the sheet of graph paper. Make a mark on the paper. Then roll the wheel across the paper through one revolution. Make a mark on the paper at that spot. Draw a line on the paper to connect the marks, which will show the distance that the wheel rolled. This distance is equal to the circumference of the wheel. (An alternative method is to wrap a cloth tape measure around the wheel.) Measure the distance and write the circumference on the Wheel Measurements table.

2. Lay the wheel flat on its side on the table. Measure the distance across the wheel through the center from one edge to the other. This is its diameter. Write the diameter on the Wheel Measurements table.

3. Divide the diameter by 2. This is the radius of the wheel. Write the radius on the Wheel Measurements table.

4. Divide the diameter into the circumference and write your answer on the table.

WHEEL MEASUREMENTS

Circumference	Diameter	Radius	Circum. Divided by Dia.

Compare your results with those of your classmates. Notice that when the circumference was divided by the diameter, the same answer (3.14) was obtained by everyone (within rounding errors), regardless of the size

NOTES

of the wheel with which he or she was working. This ratio of the diameter (*d*) to the circumference (*C*) is called pi and is represented by the Greek letter π. If $\pi = C/d$, then $C = \pi d$

5. You can use the same formula to determine diameter. Exchange wheels with another student. Measure the new wheel's circumference and solve for the diameter mathematically using the formula $d = C/\pi$:

_____

Check the accuracy of your work by actually measuring the diameter.

6. Your teacher has given you an assortment of wheels. You can find the radius (*r*) of each wheel by dividing the diameter by 2. Determine mathematically their radii starting from their circumferences.

WHEEL RADII

Wheel	Circumference	Diameter	Radius
1			
2			
3			
4			

THINKING CRITICALLY

1. Why is it important for you to know how to solve for the radius using the formula for circumference?

_____

_____

_____

2. To what distance should the legs of a compass be set in order to draw a wheel that is 20 cm in diameter?

_____

_____

3. If a wheel is 20 cm in diameter, what is its circumference?

_____

_____

CIRCUMFERENCE AND WHEEL DESIGN

It may be an advantage to have your pollution-free vehicle's wheels turn a specified number of times along the 8 meter test track. One way to accomplish this is to design a wheel with a specific circumference. The easiest way to draw a wheel is with a compass. However, you will need to know the diameter of the circle you wish to draw in order to use the compass. You can determine the diameter by solving for it mathematically. During this activity you will discover how many revolutions of the driven wheels of your vehicle are necessary in order for the vehicle to go the 8 meters specified in the Challenge.

MATERIALS AND EQUIPMENT

Quantity	Description
4	wheels of various sizes
1	calculator
1 sheet	graph paper
1	metric tape measure
1 roll	masking tape
1	marking pen

Your teacher will roll a marked wheel across an 8 meter distance. Watching the mark, count the number of revolutions the wheel makes:

_____

The track length is 8 m, or 800 cm. If the circumference of a wheel is 36 cm, then the number of revolutions required to move the length of the track is 800 cm divided by 36 cm. Thus, 22.22 revolutions would be required to go 8 meters.

Keep in mind when you design your wheels that a compass is set to the radius of the desired circle and that the radius is equal to half the diameter. For example, if you want your vehicle's wheels to make 8 turns in 8 m, then the circumference of the wheel will have to equal 1 m or 100 cm. To find out the diameter of the wheel you will need, use the same formula, $C/\pi = d$. For example:

$$100 \text{ cm}/\pi = 31.8 \text{ cm}$$

To find out the distance to which the legs of a compass or a set of dividers should be set, divide the diameter by 2. The radius of the wheel in the example above would be 31.8 cm / 2 = 15. 9 cm

Name_____ Date_____ Class _____

NOTES

1. Determine the number of revolutions necessary for the wheel size you are working on.

_____

_____

_____

2. Check your answer by actually rotating the wheel through the track distance the number of times you obtained mathematically. Why are there differences between the two values?

_____

3. Based on the data you have collected, predict the number of revolutions required for your wheel to traverse other distances. Note your answers in the table.

Distance	Number of Revolutions Required
2 meters	
4 meters	
16 meters	
32 meters	

Design details will affect the distance your vehicle travels. **Fig. 7-11** shows a possible design for a pollution-free vehicle. Note that the thread or string is wound around the axle. When the spring lever is released, it causes the string to unwind from the axle and turns the wheels. In theory, if the string is still tight after it unwinds and it was wrapped eight times around the axle, then the wheels should turn eight times. No matter what the circumference of the wheels, they will turn approximately once for every time the string is wrapped around the axle. If the string is loose after it is unwound, then the vehicle may roll farther. If the string is not attached to the axle, then it will fall free after unwinding, and the vehicle will be free to roll farther.

Suppose the string was not completely unwound from the axle even though the spring stopped moving. What changes could you make that would cause more of the string to unwind from the axle without repositioning the mousetrap or the wheel assembly?

How could this information be used in designing a vehicle that will stop on the 8 m line; no matter what type of drive you are thinking of using?

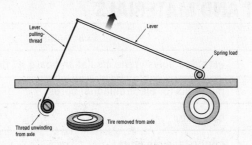

Fig. 7-11 The string unwinds from the axle.

THINKING CRITICALLY

1. How many revolutions will a wheel with a circumference of 150 cm make when rolling 8 m?

   _____

2. If your wheel turned 12 times after rolling 8 m, then what is its circumference?

   _____

3. Is there a relationship among circumference, radius, and diameter? If there is, what means could you use to communicate its pattern?

   _____

   _____

4. If an automobile tire is 60 cm in diameter, how far will the automobile travel in one revolution of the tire?

   _____

5. If the automobile in Question 4 travels 1 km, how many revolutions would the wheels turn?

   _____

   _____

RATIOS

Various combinations of pulleys could be used to build a pollution-free vehicle. During this activity, you will learn to design pulley systems. You will discover that the circumference, radius, or diameter of a wheel is related to the circumference, radius, or diameter of any other wheel to which it is connected. This relationship can be expressed as a ratio.

NOTES

EQUIPMENT AND MATERIALS

Quantity	Description
4 sq. ft.	extruded polystyrene insulation board ¾" thick
2 per class	materials for electromagnet cores
1	calculator
1	metric tape measure
1 roll	masking tape
1	marking pen
2 sq. ft	corrugated scrap cardboard
4	straight pins
120 cm	string

SAFETY FIRST

Before You Begin Make sure you understand how to handle the equipment and materials safely. Ask your teacher to demonstrate their proper use. Follow all safety rules. Refer to the Safety Handbook, beginning on page 291, for more information about safety in the lab.

1. Working with your teammates, lay out and cut wheels from the extruded polystyrene sheet using the hot wire cutter. The wheels should have the following diameters: 10 cm, 20 cm, 30 cm, and 40 cm.

2. Make a mark on the circumference of each of the wheels with a marker pen. Lay two wheels on a piece of corrugated cardboard so edges are touching. As one wheel turns, the other should turn.

3. Insert some straight pins through the centers of the two wheels. See **Fig. 7-12**. How many times do you think one wheel will turn for every revolution of the other wheel touching it?

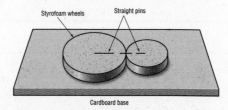

Styrofoam wheels   Straight pins

Cardboard base

Fig. 7-12 Determining the ratio between two different sized wheels

4. Check your hypothesis by rotating the wheels, carefully holding them in contact with one another while they are rotated. Enter your findings in the Wheel Revolutions table. For example, for the second column of data, rotate the 40 cm wheel one revolution and record how many revolutions the 30 cm wheel made.

Name_____ Date_____ Class_____

WHEEL REVOLUTIONS

40 cm wheel	30 cm wheel	20 cm wheel	10 cm wheel

Two wheels connected with ropes, belts, or string are pulleys. Grooves cut around their circumferences guide the ropes or belts.

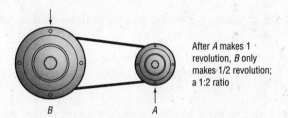

After *A* makes 1 revolution, *B* only makes 1/2 revolution; a 1:2 ratio

Fig. 7-13 Wheel or pulley ratios

5. Tie a knot in the 120 cm length of string. Then wrap the string around the wheels as shown in **Fig. 7-13** to create pulley combinations. Hold the wheels in position with the straight pins.

6. Rotate wheels while keeping the string taut. Record the number of revolutions using the wheel combinations in the Pulley Revolutions table.

PULLEY REVOLUTIONS

40 cm wheel	30 cm wheel	20 cm wheel	10 cm wheel

7. Measure the circumference and determine the circumference mathematically for each of the four wheels. Enter this information below.

WHEEL SIZES

Wheel	Diameter	Radius	Measured Circumference	Calculated Circumference
1	10 cm			
2	20 cm			
3	30 cm			
4	40 cm			

NOTES

The relationship between two wheels in a pulley can be expressed as a ratio. A ratio is a way to compare two numbers using division. For example, suppose one wheel in a pulley measures 120 cm in diameter and the other measures 20 cm. The ratio would be written 120:20. If you like the way the pulley works and want to make one with the same ratio but 5 times bigger, you simply multiply both terms by 5:

$$120 \times 5 = 600$$
$$20 \times 5 = 100$$

To make a smaller pulley—say one-tenth the size—you divide both terms:

$$120 \div 10 = 12$$
$$20 \div 10 = 2$$

It is useful to reduce ratios to their simplest form, which means dividing until there is no number except one that goes into both terms evenly. To find the simplest form of 120:20, we divide:

$$120 \div 20 = 6$$
$$20 \div 20 = 1$$

The simplest form of 120:20 is 6:1. If you wanted to re-create the same pulley arrangement, you would use a ratio of 6:1.

8. Determine the ratios among the four sizes of pulley wheels in the table below:

	10 cm dia.	20 cm dia.	30 cm dia.	40 cm dia.
10 cm dia.	10 cm			
20 cm dia.	20 cm			
30 cm dia.	30 cm			
40 cm dia.	40 cm			

9. Using information from the Wheel Sizes table (Step 7) above, determine the ratios among the circumferences of the four sizes of wheels. Enter the ratios in the table below:

Wheel	1	2	3	4
1				
2				
3				
4				

10. How do the ratios of one wheel diameter to another wheel diameter compare to the ratio of one wheel circumference to another wheel circumference?

_____

11. Would the same ratios hold true if they were among the radii of the four wheel sizes?

_____

THINKING CRITICALLY

1. A 60 cm diameter pulley is connected to another pulley with a cord. The diameter of the second pulley is unknown. When the 60 cm pulley is rotated one revolution, the other pulley rotates 4 turns. Without actually measuring the diameter of the unknown-sized pulley, what is its diameter?

_____

_____

_____

2. Pulley A is connected to pulley B. Pulley B is 12 cm in diameter and pulley A is 6 cm in diameter. For every single revolution pulley B makes, how many revolutions does pulley A make?

_____

_____

3. The ratio between the pedals and the rear wheel of a bicycle is 1:5. What does this mean in terms of revolutions?

_____

NOTES

FRICTION AND AERODYNAMICS

Engineering Technology

For this activity you will explore wheel, bearing, and body designs. See one possible design in **Fig. 7-14**. Vehicle wheels require certain characteristics, such as traction and ability to support a load. Variables affecting wheel characteristics include:

- Diameter of the wheel
- Width of the tread
- Properties of the material used to make the wheel and tread
- The mass of the wheel

The wheel must have sufficient strength to support the vehicle and withstand the forces that are applied to it during turns and when it hits obstructions.

MATERIALS AND EQUIPMENT

Quantity	Description
1 set	wheels of different materials, diameters, and widths
assorted	lubricants
assorted scrap pieces	wood, metal, and plastic
assorted	illustrations showing antique and modern cars

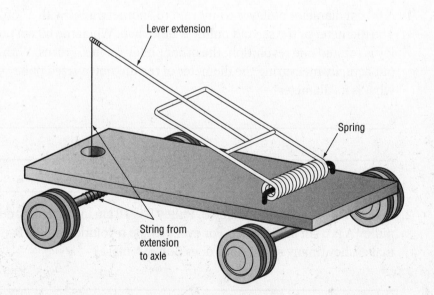

Fig. 7-14 Adding weight will load the spring more. Changing the wheel size will change the distance per revolution.

Un-powered wheels only need enough traction to guide the vehicle. Wheels that are to be powered need traction to prevent slippage when the power is applied. Traction is produced by friction. As traction increases, so does friction. However, as friction increases, an increasing amount of the energy from the power source must be used to overcome the friction. The result is a decrease in the performance of the vehicle. Wheels with narrow tread surfaces produce less friction with the road surface but provide little traction. A dragster has wide-tread tires on the rear and very narrow treads on the front for this reason.

Wheels made of different materials perform differently. Rubber may provide more traction due to its resiliency, but it produces more friction than wheels made of harder materials.

Friction between an axle and the hole in which it rotates can be reduced by proper alignment and optimal clearance. The bearings for the axles must be mounted so that the axles are perpendicular to the longitudinal axis of the vehicle. This assures that the vehicle will travel in a straight line. The bearings should produce minimal friction. Various types of lubricants can be used to reduce friction in the bearings.

Experiments with Lubricants
EQUIPMENT AND MATERIALS

Quantity	Description
1	stop watch
1	piece of lumber or hardboard, approximately 8" wide and 48" long.
1	box or stool to raise one end of the inclined plane up at approximate 30 degree angle
Assorted	Scrap pieces of wood, metal, and plastic.
Assorted lubricants	Oil, silicon, wax, etc.*

* If possible, obtain silicon in a non-aerosol container. If using it in aerosol form, the teacher should spray it on a rag and hand it to the students. Overspray from aerosol silicon can cause floors and other surfaces to become dangerously slick.

1. Set up an inclined plane about 4 feet long using a piece of lumber or hardboard. Set the inclined plane at about a 30 degree angle.

2. Place the scraps of selected materials at the top of the inclined plane. Release them and, using a stop watch, determine how long it takes for them to reach the bottom of the inclined plane. Conduct at least three trials. Depending on the smoothness of the surface of the inclined plane, the height may have to be adjusted so that all the samples will slide down the plane.

NOTES

3. Record the data in the table below.

4. Apply the selected lubricants and once again time how long it takes for them to reach the bottom of the inclined plane. Conduct three trials. Record the data.

Material	Lubricant	Time Trial 1	Time Trial 2	Time Trial 3	Average

5. You decide to retest the pieces. You notice that the pieces are sliding down the inclined plane much more quickly now. What might cause this to be the case?

_____

_____

_____

_____

Aerodynamic design is intended to reduce the friction between the body of a vehicle and the air through which it is passing. Designers must compromise between a design that is very aerodynamic and one that allows room for the power system and passengers.

THINKING CRITICALLY

1. How has air friction been reduced in the design of large trucks?

_____

_____

_____

2. What is the relationship between aerodynamic design and the consumption of fuel?

_____

_____

MASS AND ACCELERATION

NOTES

Science

During this activity you will observe how a vehicle behaves both at rest and in motion. Both are affected by mass and acceleration.

MATERIALS AND EQUIPMENT

Quantity	Description
2	bricks
1	spring scale
1 piece approx. 30 cm long	string
1 roll	cellophane tape
1	chair cart with chairs or alternative

Mass is a measure of the amount of matter an object contains. It is often equivalent to weight. Vehicle mass affects performance, both in terms of speed and accuracy in stopping at the finish line. According to Newton's Second Law of Motion, an object has inertia (the tendency to remain at rest) unless acted on by an unbalanced force. The greater the object's mass, the greater the force needed to get it moving. See **Fig. 7-15**.

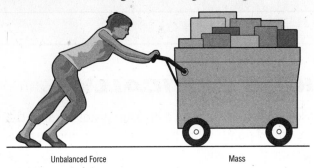
Unbalanced Force   Mass

Fig. 7-15 The greater the object's mass, the greater the force required to get it moving.

1. Connect a brick to a spring scale with a piece of string. Let the brick represent a pollution-free vehicle. Pull on the spring scale and move the brick about 40 cm. How did the readings on the scale change?

_____

_____

_____

2. The friction between the brick and the table increases over a range until it is less than the force being applied to the brick through the spring scale. Repeat with a second brick placed on top of the first.

Name_____ Date_____ Class _____

NOTES

Less mass means it is easier to get a vehicle moving and increase acceleration. The vehicle's **acceleration** is its increasing forward movement. However, less mass also means there is less traction between tire and road, so there is a tradeoff to decreasing weight. This is why people add weight to a car to increase traction in snowy winter weather.

Force is measured in newtons or pounds. Force is the product of mass and acceleration (force = mass x acceleration). If the force supplied by an energy source is constant, then as the mass increases, the acceleration decreases. Likewise, if the mass is decreased, the acceleration will increase, given a constant force.

SAFETY FIRST

Before You Begin Use caution during this activity. A loaded, moving cart can cause serious injury to students and/or physical to the facility.

3. A volunteer should push a loaded cart as hard as he or she can while the class observes the acceleration. Experiment with changing the force (e.g., using two students to push) and changing the mass (e.g., removing half of the chairs). Also, experiment with the force required to bring the cart to a stop under different loads. What do you observe?

_____

_____

THINKING CRITICALLY

1. Suppose some students used a braking system to stop their vehicles at the specified point. What happened to the energy that was used to actually stop the vehicle?

_____

2. What would be an advantage and a disadvantage to building vehicle momentum?

_____

_____

DESIGNING YOUR POLLUTION-FREE VEHICLE

Engineering Technology

During this activity, you will use your notes and sketches to design your pollution-free vehicle. One possible design is shown in **Fig. 7-16.** You will make more sketches and a three-view working drawing, showing the front, top, and right side.

MATERIALS AND EQUIPMENT

Quantity	Description
assorted	materials about transportation systems
assorted	non-polluting power units
3 sheets	plain paper
1	pencil
1	architect's scale

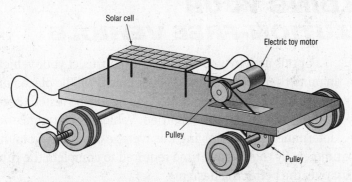

Fig. 7-16 The string winds up to stop the vehicle after 8 meters.

1. Divide your design task into two phases: the power system and the chassis system. Consider and apply the principles you have been studying. Draw heavily upon research.

2. Prepare at least three sketches of possible power systems, and three sketches of possible chassis designs.

3. Select the designs that you think will work best.

4. Check your final design against the Design Brief.

5. Create a finished three-view drawing of your chosen design.

6. Keep all sketches and drawings to turn in at the end of the project.

THINKING CRITICALLY

1. What science and math principles did you apply?

_____

_____

_____

2. Do you think that another person could build your design if you gave them your sketches and drawings? Why or why not?

_____

NOTES

Modeling

During this part of the project you will use a wide variety of tools and materials to produce the vehicle and energy transmission system that you have designed. You will do a lot of informal testing using the 8-meter track that will later be used in determining a winner.

BUILDING YOUR POLLUTION-FREE VEHICLE

For this activity you will actually construct your vehicle. Keep in mind that during the test your vehicle will be placed on a pre-determined starting line. Each vehicle will be tested three times, with the results of each recorded. Vehicle performance will be determined by multiplying the average deviation from the finish line in centimeters by the average time required to complete the run. The lower this value, the better the design.

MATERIALS AND EQUIPMENT

Quantity	Description
1 set	hole saw and cutting wheels for scroll saw or band saw
1	portable electric drill or drill press
assorted	scrap wood, metal, and plastic
1	scratch awl
1	dividers for sheet metal
1	compass with pencil
1	centering head from combination square
1	bench rule
1	duck bill or aviation snips for sheet metal

SAFETY FIRST

Before You Begin Make sure you understand how to use the tools and materials safely. Ask your teacher to demonstrate their proper use. If you use the hole saw, be sure the material is securely clamped to the work table. Follow all safety rules and wear eye protection. Refer to the Safety Handbook, beginning on page 291, for more information about safety in the lab.

Project 7 Modeling

NOTES

1. Use the hole saw to cut out wheels for your vehicle. It allows you to locate the axle hole in the center of the wheel automatically. Choose the size cutter you need, use the scratch awl to center punch, and drill out the wheels.

2. Use the band saw for cutting larger wheels. First lay out the circumference of a wheel with a compass, being careful to reduce wasted material. The depression made by the point of the compass will help you place the drill for drilling the hole for the axle. Make relief cuts in the stock so the blade does not bind and so the blade can be easily backed out. Use the guard and never place fingers in direct line with the blade.

3. Use the centering head of the combination square as needed to locate the center of a circular piece of stock.

4. Lay out circles on sheet metal using the scratch awl and dividers to scribe layout lines. Use snips for cutting circles from sheet metal. Thin, semi-flexible sheet plastic such as polystyrene can also be cut with sheet metal snips.

5. Your teacher may introduce you to other techniques and materials that can be used for making wheels, depending upon the equipment that is available.

6. Follow your three-view drawing to finish building the other parts of your vehicle.

7. Check your vehicle against the specifications in the Design Brief to be sure it meets them all.

8. Test your vehicle on the test track.

9. Make any needed adjustments.

THINKING CRITICALLY

1. What changes did you make to your original design as you constructed your solution? Why did you make these changes?

_____

_____

_____

2. What did you do with your design to make your vehicle more aerodynamic?

_____

_____

NOTES

DETERMINING THE COEFFICIENT OF FRICTION

For this activity you will have a chance to improve your vehicle before it attempts to meet the challenge. You will be able alter the effects friction will have on it.

MATERIALS AND EQUIPMENT

Quantity	Description
1	pulley
1 length	string
assorted	weights

As the weight of the vehicle increases, the frictional forces opposing its motion increase. The value used to compare the amount of friction to a given weight is called the **coefficient of friction** (μ). Since a vehicle moving at a constant velocity has no net or total force pushing or pulling it, the frictional forces are equal in magnitude to the applied forces.

1. Set up the system shown in **Fig. 7-17**. Add weight to the hanging string until a slight push causes the vehicle to move at a constant velocity. It should neither speed up nor slow down.

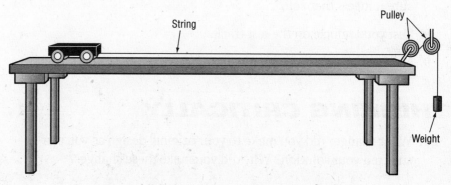

Fig. 7-17 Determining the coefficient of friction

2. Record the amount of weight hanging on the string.

_____

_____

N O T E S

This is the force equal in magnitude to the frictional force that is resist-ing the motion of the vehicle. This weight divided by the weight of the vehicle gives the coefficient of friction:

$$F/W = \mu$$

F is the force of friction, W is the weight of the vehicle, and μ is the co-efficient of friction.

3. Change the bearings and/or wheels on your vehicle and repeat the experiment.

4. Record the amount of weight hanging on the string.

_____

5. Compare the results of the two tests.

THINKING CRITICALLY

1. How does adding weight to a vehicle affect friction?

_____

_____

2. Is it possible to add too much weight to a car when traveling in snow?

_____

_____

_____

_____

_____

3. What are the sources of undesirable friction in wheeled vehicles?

_____

_____

_____

_____

_____

_____

Project 7 Modeling

Evaluation

During the evaluation part of this project, you will formally test your vehicle by racing it in a series of three timed trials on the 8 meter track. After each trial, you will be allowed to rework your design, so as to improve its performance.

TESTING YOUR POLLUTION-FREE VEHICLE

During this activity, you will determine how well your vehicle is performing and what design variables are causing a good or poor performance.

MATERIALS AND EQUIPMENT

Quantity	Description
1	metric tape measure
Assorted	stopwatch

1. Set up the starting line and the finish target 8 meters apart on a smooth floor surface. Use a stopwatch to time the three trials and a tape measure to determine the deviation from the target at the finish line.

2. Record your vehicle's performance data on the Vehicle Performance table. Enter only the elapsed time, the deviation value, and changes made. The Challenge called for your vehicle to stop at a specified target. The deviation is how far from that target it actually stopped. The other values will be filled in later.

VEHICLE PERFORMANCE

Trial	Elapsed Time in Sec.	Deviation in cm	Average Speed in cm/sec	Performance Index	Changes Made
1					
2					
3					
Average					

3. Test your vehicle three times. After each trial, modify your vehicle to improve performance and write a brief description of the changes you made in the performance table. Use an extra sheet of paper.

4. Determine vehicle performance by multiplying the average deviation from the finish line in centimeters by the average time required to complete the run. The lower this value, the better the design.

THINKING CRITICALLY

1. Are the wheels you used on your vehicle realistic in scale? What are advantages and disadvantages of large wheels on real autos?

_____

_____

2. Discuss performance of heavy vehicles versus light vehicles in terms of speed and accuracy of stopping at the finish line?

_____

_____

ANALYZING VEHICLE PERFORMANCE

During this activity you will analyze your vehicle's performance. You may also redesign the vehicle drive and wheels.

MATERIALS AND EQUIPMENT

Quantity	Description
1	calculator
1	metric tape measure
1	stopwatch
1 roll	masking tape
3 sheets	graph paper

1. Calculate the following values and enter them on the Vehicle Performance Table that you began in the last activity:
   - The average elapsed time in seconds.
   - The average deviation across the three trials.
   - The average speed for each of the three trials in meters per second. (The length of the track was 8 meters.)

NOTES

2. Calculate the Performance Index for each of the vehicles. This is the product of the average elapsed time and the average deviation.

3. As a class, record the Performance Index for each student on a chart that can be displayed for the entire class.

4. Calculate the circumference of the wheels to which power was applied. Then, using the data from the Vehicle Performance table, determine the number of revolutions that your vehicle's wheels turned from start line to finish line (track length was 8 meters).

_____

5. Calculate revolutions per second using the elapsed time that was recorded in the Vehicle Performance table.

_____

6. Determine the following:
   • The ratio of the tire diameter to the axle diameter.

   _____

   • The ratio of the circumference of the tire to the circumference of the axle.

   _____

   • The ratio of the diameter of the rear wheels to the diameter of the front wheels.

   _____

THINKING CRITICALLY

1. How could you determine if a tire on a vehicle slipped if you knew the diameter of the tire and how many revolutions the tire turned in one kilometer?

   _____

   _____

   _____

   _____

2. As tires wear, will the speedometer show a speed that is lower or higher than when the tires were new? Why?

   _____

   _____

   _____

   _____

Hydroponic System

Design Brief

Technology has changed agriculture in America. The small farms worked with horse-drawn plows have given way to large farms managed with heavy equipment and computers. In addition, technology has helped to develop **hydroponic farming**—growing plants without any soil. See **Fig. 8-1**. Hydroponic farming can be efficient and cost effective, particularly where good soil is scarce as in desert areas. One place where hydroponic farming may someday be the only possible method for providing food is outer space. If and when people colonize a permanent space station, they will probably have a hydroponic farm. Meanwhile, hydroponic vegetables and other plants are currently grown here on Earth. For example, almost all of the fresh tomatoes sold in your local grocery store are hydroponically grown.

Fig 8-2 This greenhouse provides space for growing plants without soil in a hydroponic garden.

Challenge

Design and construct an efficient hydroponic system, plant and maintain a crop, and produce maximum total plant growth.

Criteria

- Your hydroponic system must be able to accommodate 12 plantings grown from the same type of seeds.
- It must automatically provide nutrients and air to the plants as needed.
- You and your team must maintain the health of the crop, thinning it after 7–10 days to 6 plants.
- You must measure and record plant growth, nutrient pH, and other data at regular intervals during a one-month period.

NOTES
- At the end of one month, you will calculate total plant growth to determine the winner of the challenge.
- All designs must be accompanied by appropriate documentation, including:
  - ☐ Sketches of all possible solutions considered
  - ☐ A final drawing of your chosen solution
  - ☐ Data collected from various tests and experiments
  - ☐ A chart/graph showing how your solution performed
  - ☐ A cost analysis for operating 1,000 of your hydroponic systems
  - ☐ Information gathered from resources
  - ☐ Notes made along the way

Constraints

- Your system must not contain any soil.
- It cannot exceed a volume of 3,000 cu cm (about 1 cu. ft.).

Engineering Design Process

1. **Define the problem.** Write a statement that describes the problem to solve. The Design Brief and Challenge provide information.

2. **Brainstorm, research, and generate ideas.** The Research & Design section, beginning on page 241, provides background information and activities that will help with your research.

3. **Identify criteria and specify constraints.** Refer to the Criteria and Constraints on pages 239 and 240. Add any your teacher recommends.

4. **Develop and propose designs and choose among alternative solutions.** Remember to save all the proposed designs so that you can turn them in later.

5. **Implement the proposed solution.** Once you have chosen a solution, determine the processes you will use to make your hydroponic system. Gather the tools and materials you will need. Make sure you understand and follow all safety rules.

6. **Make a model or prototype.** The Modeling section, beginning on page 260, provides instructions for building the hydroponic system.

7. **Evaluate the solution and its consequences.** The Evaluation section, beginning on page 262, describes how to test the system.

8. **Refine the design.** If instructed by your teacher, make changes to improve your design.

9. **Create the final design.** If instructed by your teacher, make a new hydroponic system based on the revised design.

10. **Communicate the processes and results.** Write a report about your project. Turn in all the documentation listed under Criteria.

Research & Design

Before designing your solution, you and your teammates will study hydroponic systems (technology), plant structures and nutrients (science), and nutrient ratios (mathematics). They will give you the information you need to take the Challenge.

GROWING PLANTS WITHOUT SOIL

Engineering Technology

Soil provides nutrients, water, and support for plants. It also allows a small mount of air to reach their roots. In hydroponic systems, where no soil is used, food and water are delivered by a nutrient solution. There is no one "best" nutrient solution that works for all plants. Different plants and growing conditions require different formulas. In addition, plants need light, warmth, carbon dioxide, and oxygen. Plants take in oxygen through their leaves above ground, but the roots need to take in oxygen also. In traditional soil-based planting, the soil, if not too packed down, allows air to reach the roots. In hydroponic systems, oxygen is often pumped into the nutrient solution.

MATERIALS AND EQUIPMENT

Quantity	Description
assorted	literature about basic hydroponic farming

Humans, too, cannot live under water without a supply of air. How do divers breathe underwater?

_____

_____

A wide range of commercial and experimental hydroponic systems has been developed to provide all the essentials plants need. The basic methods of hydroponic farming are water culture, aggregate culture, and aeroponics.

Water Culture Systems

In the water culture method, plants are suspended in water that contains the nutrients they need. See **Fig. 8-2** on page 242. One commercial water culture system is the "float bed" system in which plants float above the nutrient solution on a Styrofoam™ support. Leafy vegetables such as lettuce and spinach grow well in water culture systems.

NOTES

Water culture systems may be open or closed. Fig. 8-2 shows a closed system. Closed systems are fairly simple to set up, but they may not be as effective as those that provide for fresh supplies of air and nutrients. Plants may also be supported by an inert substance such as sawdust or peat suspended on a mesh of plastic, wire, or cheesecloth. Oxygen is generally pumped into the nutrient solution. See **Fig. 8-3**.

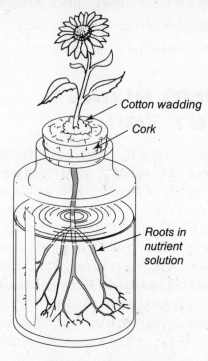

Cotton wadding

Cork

Roots in nutrient solution

Fig. 8-2 The cork and cotton support the plant in this variation of the closed water culture system.

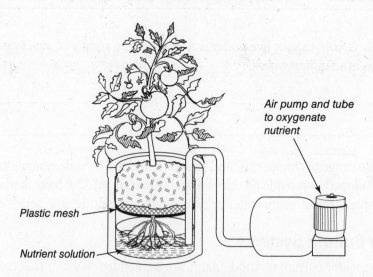

Air pump and tube to oxygenate nutrient

Plastic mesh

Nutrient solution

Fig. 8-3 A water culture hydroponic system

Project 8 Research & Design

Aggregate Culture Systems

   In the aggregate culture method, the roots of the plant are anchored in an inert material such as sand, gravel, or vermiculite. See **Fig. 8-4**. The nutrient solution is absorbed into or circulated through the aggregate. Spaces among the particles of aggregate allow air to reach the roots of the plants. One commercial aggregate culture system is the ebb and flow, or flood and drain, system. Plants are placed in an aggregate bed and the bed is flooded with nutrient solution. The solution is then drained or pumped out, allowing the roots to receive the oxygen they need. The ebb and flow cycle is regulated to produce the maximum yield for a given crop.

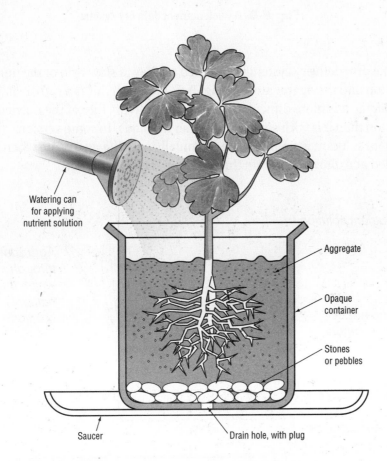

Watering can for applying nutrient solution

Aggregate

Opaque container

Stones or pebbles

Saucer

Drain hole, with plug

Fig. 8-4 In aggregate culture, the roots of the plant rest in the growing medium, such as perlite, vermiculite, rock wool, or sand/gravel.

   In one type of aggregate culture system, a wick runs from a reservoir containing the nutrient solution to the aggregate material that supports the roots. See **Fig. 8-5** on page 244. Capillary action draws the nutrient solution from the reservoir to the aggregate. The size and type of the aggregate particles affect the balance between air and nutrient solution.

NOTES

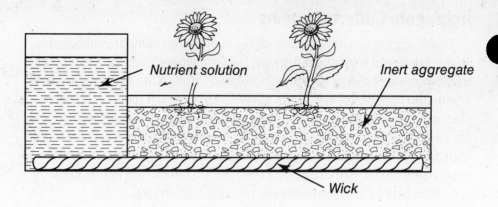

Fig. 8-5 A wick nutrient delivery system

The drip delivery system is designed to allow a slow drip of the nutrient solution into the aggregate. See **Fig. 8-6**. Although this is a rather simple technique, it is more difficult to control the delivery rate of the nutrient solution than it is with some of the other methods. In some versions, the nutrient solution is simply poured manually over the aggregate when needed and allowed to drain through as in the drip method.

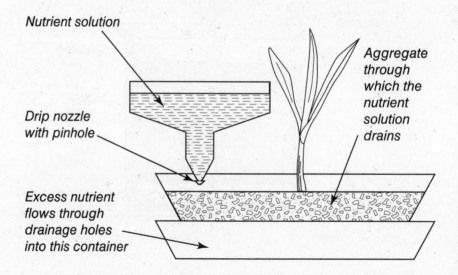

Fig. 8-6 A drip nutrient delivery system

Aeroponic Systems

A third system, called aeroponics, is also used. In aeroponics, the roots are suspended in air. Either a thin mist of nutrient solution surrounds the roots at all times, or the roots are sprayed periodically. See **Fig. 8-7**.

Name_____ Date_____ Class_____

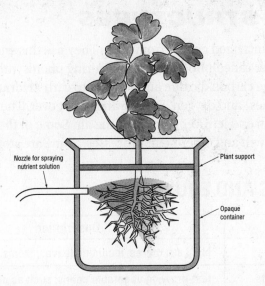

Nozzle for spraying
nutrient solution

Plant support

Opaque
container

Fig. 8-7 An aeroponic system uses less nutrient solution
than water culture uses.

THINKING CRITICALLY

1. How might you describe a hydroponic system in terms of the
universal systems model?

_____

_____

_____

_____

2. Would hydroponically grown vegetables taste, smell, look, or feel
differently from conventionally grown vegetables? Explain.

_____

_____

_____

3. What might be some of the advantages and disadvantages of hydroponic crop production?

_____

_____

_____

_____

NOTES

PLANT STRUCTURES

Plants need energy from light. They use this energy in a process called **photosynthesis**. During photosynthesis, they change carbon dioxide and water into carbohydrates (sugars and starches) and oxygen. Some water is left over. They release the oxygen and leftover water into the air. Some of the carbohydrates are used for growth and other plant processes. Some are stored.

MATERIALS AND EQUIPMENT

Quantity	Description
assorted	plant structures, including leaves, stems, and roots
1 or 2 packets	fast-growing vegetable seeds, such as pea and radish
1	container for seed germination, such as Petri dish
1	hand lens
handful	cotton
1	compound microscope
assorted	slides showing longitudinal and transverse cross sections of roots and root tips

When plants live outside, they obtain the light they need from the sun. When hydroponic systems are located inside, the plants are usually provided with artificial light. Plants obtain carbon dioxide from the air. It enters the plant through small openings in the leaves called stomata. See **Fig. 8-8**. (A single opening is called a stoma.)

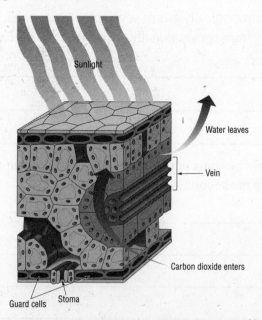

Sunlight

Water leaves

Vein

Carbon dioxide enters

Guard cells   Stoma

Fig. 8-8 A cross section of a leaf: Guard cells around the stoma control the size of the opening.

Plants absorb water through their roots. Much of this water is released later through the plant's leaves and must be replaced. Plants that grow outdoors obtain water from rain and humidity in the air. This helps limit which plants will thrive in a particular climate. When plants live indoors in a controlled environment, it is possible to grow many different species.

NOTES

1. Examine a leaf using the hand lens and microscope. Attempt to see the leaf's stomata.

2. Examine roots from two different plants. How are they similar? How are they different?

_____

_____

_____

3. How do the structures you see relate to where the parts are found?

_____

_____

_____

_____

_____

_____

4. Place several seeds in various positions in a Petri dish stuffed with cotton saturated with water. Set them aside until the seeds sprout.

5. Observe the direction in which the roots grow. Describe what you see.

_____

_____

6. Move the dish to a location that has different temperature and light. Leave it there for several days. Describe what you observe.

_____

7. Using a compound microscope, examine a prepared slide of a longitudinal cross section of a root tip. How are the cells at the tip of the root different from the other cells?

_____

_____

_____

NOTES

8. What does the tip of the root do? Why might it be important to have special cells in the root tip?

_____

9. How do cells seem to change as you move from one end of the root to another? Where do you think the newest cells are? Where are the oldest cells? What evidence can you cite to support your answer?

_____

_____

_____

_____

_____

10. Using the compound microscope, observe a prepared slide of a transverse cross section of the mature region of a root. You will see a large array of cells. Look for patterns and changes. Remember that the function of roots is to take in water and nutrients and deliver them to the other parts of the plant. What do you see that relates to this known function?

_____

_____

THINKING CRITICALLY

1. Why didn't some of the seeds you planted germinate?

_____

_____

_____

_____

2. What advantage to the plant comes from producing root hairs?

_____

_____

3. What factors would make growth more difficult for a plant's roots?

_____

_____

THE PH SCALE

The pH value of a substance is a measure of how acid or alkaline (basic) it is. See **Fig. 8-9**. Vinegar and lemon juice, for example, are acid. Laundry detergent is alkaline. Alkaline substances are called bases.

MATERIALS AND EQUIPMENT

Quantity	Description
1 roll	litmus paper tape
3	small containers
1 liter	nutrient solution
1 bottle	white vinegar
1 box	baking soda
approx. 4 liters	tap water
assorted	darticles about acid rain

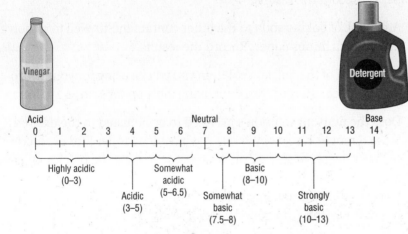

Fig. 8-9 The pH scale has values from 0 to 14, with values below 7 being acidic and above 7 being basic. Each number has a factor of 10.

Why is pH value important? If the pH value of the soil or nutrient solution is too acid or too alkaline, the plants cannot use the nutrients. Although there are exceptions, most plants do best when the soil or nutrient solution is slightly acidic (5.6-6.5), as shown in the list below.

pH Preferences of Common Vegetables and Herbs

Broccoli 6.0–7.5	Cucumber 5.5–7.0	Parsley 5.0–7.0
Carrot 6.0–7.0	Garlic 5.5–8.0	Pole bean 6.0–7.5
Celery 5.8–7.0	Lettuce 6.0–7.0	Radish 6.0–7.0
Chives 6.0–7.0	Mustard 6.0–7.5	Spinach 6.0–7.5

NOTES

Litmus paper is used to test for pH value. Its color after being dipped into a solution indicates the relative pH of the solution.

Indicator Color	Relative pH
Bright red	Strong acid
Red	Medium acid
Reddish purple	Weak acid
Purple	Neutral
Blue green	Weak base
Green	Medium base
Yellow	Strong base

1. Place about 125 mL water in each of two containers. Use a piece of litmus paper to test the pH of one of the containers of water. The litmus paper will change color based on how acid or alkaline the water is. Record the results in the pH Test Table.

2. Add 5 mL of vinegar to one of the containers. Stir. Test with litmus paper. Record the results.

3. Add 5 mL of baking soda to the other container. Stir well to dissolve it. Test with litmus paper. Record the results.

4. Test the pH of the nutrient solution you will be using for your hydroponics system. Record the result in the table under Nutrient solution 1.

5. Does the nutrient solution's pH need to be adjusted to the desired 5.6-6.5 range? If so, try 2½ mL of white vinegar per liter of solution to lower the pH or 2½ mL of baking soda to raise the pH. It may take a few hours for the solution to adjust, so it's better to wait a while before testing the pH. Record the results under Nutrient solution 2. Indicate whether you added vinegar or soda and how much.

TESTING FOR PH VALUE

Substance	Color of Litmus	Relative pH
Tap water		
Water and vinegar		
Water and baking soda		
Nutrient solution 1		
Nutrient solution 2		
Substance added:	Amount:	

THINKING CRITICALLY

1. What do you think will happen if a plant receives a nutrient solution that has too high a pH level? Too low a pH?

    _____

    _____

    _____

2. Look up the word *hydroponics* in a dictionary that includes word origins. What are those words and what do they mean? Why do you think they were chosen to describe farming without soil?

    _____

    _____

NUTRIENTS AND PLANT GROWTH

Science

Plants obtain most of the things they need from the air and water. The remainder comes from chemical nutrients found in the soil. The most important of these nutrients are nitrogen, phosphorous, and potassium.

Although these nutrients are found in the soil, the soil in different regions may have differing amounts. Also, if the land has been farmed regularly, levels of some nutrients may be low. In that case, farmers test the soil to find out what is needed and then add fertilizers. There are many nutrient formulas that will work. In hydroponic systems, the nutrients are added to water to create the nutrient solution.

MATERIALS AND EQUIPMENT

Quantity	Description
assorted	labels from commercial fertilizers and nutrient solution
1	38-46 L (10-12 gal.) container for mixing and storing nutrient solution
1	metric scale
1 set	measuring spoons
1	container for mixing dry ingredients
1	liter bottle
several grams	plant nutrients
several milliliters	trace elements

Name_____ Date_____ Class _____

1. Your teacher will distribute labels from commercial fertilizer and/or nutrient solution mixes. As a class, make a list of ingredients on the labels. How many different nutrients are used?

_____

_____

_____

_____

2. Why do you think some plants have different requirements than others?

_____

_____

3. How might plant nutrition and human nutrition be alike and differ?

_____

_____

_____

4. As a class, use this basic nutrient formula to mix a nutrient solution:

Step 1. Mix the following salts with 38 L (10 gal.) water.

Salts	Quantity	Nutrients Supplied
Sodium nitrate	30 grams	Nitrogen
Potassium sulfate	10 grams	Potassium, sulfur
Superphosphate	12 grams	Phosphorus, calcium
Magnesium sulfate	8 grams	Magnesium, sulfur

Step 2. Mix the following trace elements and store in dry form.

Trace Elements	Quantity	Nutrients Supplied
Ferrous sulfate	30 mL (1 oz.)	Iron
Manganese sulfate	5 mL (1 tsp.)	Manganese
Boric acid powder	5 mL (1 tsp.)	Boron
Zinc sulfate	2.5 mL (½ tsp.)	Zinc
Copper sulfate	2.5 mL (½ tsp.)	Copper

Step 3. Dissolve 2.5 mL (½ mL tsp.) dry trace element mixture in 1 L water. Add 82 mL of this solution to each 38 L of the solution mixed in Step 1. Discard any unused portion of the trace element solution because it will not keep.

5. What might happen if you were to double the amount of nutrients available?

_____

_____

THINKING CRITICALLY

1. Where do the nutrients found in soil come from?

_____

_____

2. Why are fertilizers necessary for some lawns?

_____

_____

_____

_____

NUTRIENT RATIOS

The relationship among ingredients in a nutrient formula can be expressed as a ratio. A ratio is a way to compare two numbers using division. For example, suppose a mixture you want to try requires 120 grams of sodium nitrate and 20 grams of magnesium sulfate. The ratio would be written 120:20. If you like the results of using the mixture and want to make 5 times as much, you simply multiply both terms by 5:

$$120 \times 5 = 600$$
$$20 \times 5 = 100$$

To make a smaller amount—one-tenth the size—divide both terms:

$$120 \div 10 = 12$$
$$20 \div 10 = 2$$

MATERIALS AND EQUIPMENT

Quantity	Description
1	calculator

NOTES

It is useful to reduce ratios to their simplest form, which means dividing until there is no number except one that goes into both terms evenly. To find the simplest form of 120:20, we divide:

$$120 \div 20 = 6$$
$$20 \div 20 = 1$$

The simplest form of 120:20 is 6:1. If you wanted to re-create the same mixture, you would use a ratio of 6:1.

1. The HydroGreen Plant Food Analysis table shows the ratio of the different ingredients in the HydroGreen Hydroponic Plant Food. At the top of the table is a ratio—10:8:22. To what does the ratio refer?

    _____

    _____

HYDROGREEN PLANT FOOD ANALYSIS

Guaranteed Analysis—10:8:22		
Total Nitrogen		10%
Nitrate Nitrogen	8%	
Ammoniacal Nitrogen	2%	
Available Phosphoric Acid		8%
Soluble Potash		22%
Calcium	5%	
Magnesium	1%	
Sulfur	3%	

2. If one of the compounds in the HydroGreen Plant Food is increased, what must happen to the quantities of the other compounds in order to maintain the same ratio?

    _____

    _____

3. By looking at the analysis of ingredients on the plant food label, you know the *ratio* of the components. Do you necessarily know the actual *quantities* that are in the solution? Why or why not?

    _____

    _____

    _____

4. Below is a partial list of ingredients found in the Basic Nutrient Formula you worked with in the previous activity. Determine the ratio.

NOTES

_____

Salts	Quantity
Sodium nitrate	30 grams
Potassium sulfate	10 grams
Superphosphate	12 grams
Magnesium sulfate	8 grams

THINKING CRITICALLY

1. What does it mean for the compounds to be in equal proportion? For example, is there any difference in the ratio 1:1:1 and 2:2:2?

_____

_____

2. Suppose you need a mixture that has the ratio A:B:C = 3:1:4, and you have only 2 g of compound B available. What amounts of the other two compounds are needed?

_____

_____

_____

VOLUME OF HYDROPONIC CONTAINERS

You will need to determine the volume of your hydroponic container to be sure it falls within the 3,000 cubic centimeter design constraint. Your teacher will provide a collection of containers in various shapes, such as prisms, cylinders, cones, and pyramids. See **Fig. 8-10** on page 256. How many different categories can you identify?

NOTES

MATERIALS AND EQUIPMENT

Quantity	Description
assorted	hydroponic containers
assorted	containers that are related in shape or form
Several	tape measures
Sample for determining volume	water or rice
several	measuring cups or graduated beakers
1 dozen	2″ × 3″ × 4″ blocks cut from 2 × 4s
1 sheet	¼″ or ½″ graph paper

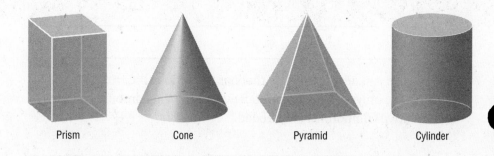

Prism   Cone   Pyramid   Cylinder

Fig. 8-10 Related geometic forms have an identical base and height.

You will need to determine the volume of your hydroponic container to be sure it falls within the 3,000 cubic centimeter design constraint. Your teacher will provide a collection of containers in various shapes, such as prisms, cylinders, cones, and pyramids. See Fig. 8-10. How many different categories can you identify? How are the shapes of the containers alike? How are they different?

Think about container volumes in terms of the heights and the area of the bases of the containers rather than as a list of formulas specific to each shape. For example, list 10 pairs of numbers that have a product of 80. Think of the two numbers as a base area and a height of a shape with a volume of 80 cubic units. How are they related? If you said, "as one goes up, the other goes down," you are correct. This set of numbers can be graphed to depict this relationship visually.

1. Examine the containers. Use graph paper to wrap small, box-like containers or blocks of wood to determine the surface area.

NOTES

2. Fill the prisms and "related" cylinders with water or rice and record the volume in milliliters: _____ (Remember that, for water, 1 cc = 1 mL.) Determine the area of the base and the height of the shapes using centimeters.

_____

_____

3. Develop the mathematical models (formulas, equations) for the volumes of the cylinder and the rectangular prism.

_____

_____

_____

_____

4. Fill the prisms and "related" pyramids with water or rice. What do you think is the relation between the volume of the prism and its related pyramid? Develop the mathematical model for the volume of a pyramid.

_____

_____

5. Fill the cylinders and "related" cones with water or rice. What do you think is the relation between the volume of a cylinder and its related cone? Develop the mathematical model for the volume of a cone.

_____

_____

6. What measurements are necessary to determine the surface area and volume of each of the container shapes? Explain your answers.

_____

_____

_____

_____

NOTES

THINKING CRITICALLY

1. How is the area of the base of a container shape related to its volume if the height is constant?

_____

_____

_____

2. How is the height of a container shape related to its volume if the base area remains constant?

_____

DESIGNING YOUR HYDROPONIC SYSTEM

Engineering Technology

Consider what you have learned so far, do additional research if you wish, and design your hydroponic system. Decide whether to use water or aggregate culture, how to deliver the nutrient solution, and how to provide oxygen to plant roots.

MATERIALS AND EQUIPMENT

Quantity	Description
several sheets	sketching and drawing paper
several	drawing pencils
1	computer and CAD or other drawing software (optional)

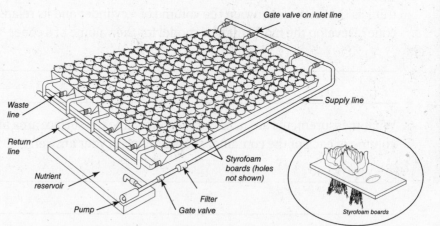

Fig. 8-11 In this raceway system, the Styrofoam boards suspend the body (stems and leaves) of the plant over the nutrient solution. The roots hang in the solution.

NOTES

One possible design is shown in **Fig. 8-11**. It uses a method commonly found in commercial systems. A pump circulates nutrients through "raceways" of plastic tubes. The plants are suspended on Styrofoam boards that float in troughs. Their roots float in the nutrient solution.

Another design is shown in **Fig. 8-12**. In this system, black polyethylene plastic is stapled around the seedlings. The nutrient solution is pumped continuously through the channels formed by stapled plastic.

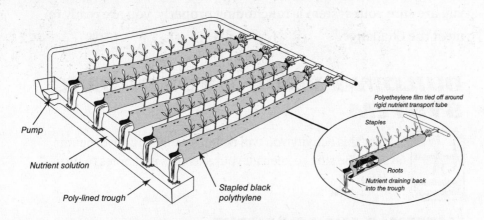

Fig. 8-12 The nutrient film technique pumps nutrient continuously through tubes of polyethylene film wrapped around seedlings.

1. Think about what you have learned about hydroponic systems. Sketch at least three possible designs for your system.

2. Select the system design that you think will work best.

3. Check your chosen design against the Criteria and Constraints listed in the Design Brief on page 239. Make any necessary changes.

5. Make a finished drawing. If available, use drawing software.

THINKING CRITICALLY

1. What science and/or mathematics principles did you apply as you went about designing your solution?

_____

_____

2. Predict mathematically the volume of the container you intend to use.

_____

_____

Modeling

For this part of the project, you will use a variety of tools, materials, and equipment to construct the hydroponic system you designed. You will have an opportunity to work out the "bugs" in your design. After you are sure your system is functioning properly, you are ready to meet the Challenge.

BUILDING YOUR HYDROPONIC SYSTEM

Engineering Technology

For this activity you will actually build your hydroponic system. Be sure to keep all your sketches and notes to turn in later.

MATERIALS AND EQUIPMENT

Quantity	Description
assorted	recycled materials, such as plastic bottles, egg cartons, soda cans
assorted	wood, metal, and plastic materials
assorted	fasteners and hardware
approx. 1220 mm	⅛″ plastic tubing
several	fittings for tubing
1	hot glue gun and glue
assorted	assorted actuators, aquarium air pumps, water pumps, motors

SAFETY FIRST

Before You Begin Make sure you understand how to use the tools and materials safely. Ask your teacher to demonstrate their proper use. Follow all safety rules and wear eye protection. Refer to the Safety Handbook, beginning on page 291, for more information about safety in the lab.

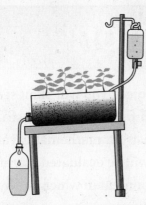

Fig. 8-13 The more effective your design, the better your plants will grow.

1. Build your hydroponic system based on your final drawings. See **Fig. 8-13**.

2. Test your system's components to be sure they work as planned.

3. Check your system against the Criteria and Constraints listed in the Design Brief on page 239.

4. In the previous activity, you predicted the volume of your container. Measure its volume using water or rice. Was your prediction correct?

_____

5. Observe the systems other teams have constructed. Which designs do you think will be most effective?

_____

_____

THINKING CRITICALLY

1. As you constructed your solution, what changes did you make to your original design? Explain why you made these changes.

_____

_____

_____

2. If you were to start over and design a completely new solution, what would you do differently?

_____

_____

_____

Project 8 Modeling

NOTES

Evaluation

During this part of the project, you will plant seeds, allow them to grow, and monitor their progress. You will use the data you collect during this time as the basis for mathematical analyses. At the end of one month, your results will be evaluated. Careful attention must be paid to the general environment in which the hydroponic systems will be set up to ensure conditions are the same for all teams.

CONTROLLING THE GROWING ENVIRONMENT

Engineering Technology

Place your hydroponic system where conditions are right for plant growth. Lighting needs vary with the plants. Fruits and vegetables require high levels of light energy, while herbs and lettuce can be grown with lower levels. Sunlight through a south or west window may be sufficient for herbs and lettuce, but supplementary artificial lights will be needed for fruits and vegetables.

MATERIALS AND EQUIPMENT

Quantity	Description
4 or more	dual-bulb, 40 W fluorescent "shop light" fixtures and cool white bulbs, 4 ft. long
2 per fluorescent fixture	15 W incandescent bulbs
1	electric switch and wire
assorted	materials for constructing support for lights
1	electric timer (optional)
1 or more	thermometer

If classroom space that receives enough natural sunlight is not available, your class may need to build an artificial lighting system. This can be done with a series of fluorescent light bulbs suspended above the plants. Different plants may require different colors of light. Four different types and colors of fluorescent tubes are available: cool white (strong blue, medium red), warm white (medium blue, medium red), plant tubes (strong blue, strong red), and full spectrum bulbs. However, researchers have found that 40-watt cool white tubes in combination with 15-watt incandescent bulbs tend to provide a good range of artificial light for hydroponic systems.

If artificial lighting will be needed, lighting and temperature should be alternated to simulate day and night conditions for the plants. The fluorescent light bulbs should be kept several inches above the plants and be left on for about 16 hours each day. The temperature should be about 22°C (72°F) during the day and about 16°C (62°F) at night.

1. Determine the amount and the color of light that your particular plants will need.

2. If artificial light is necessary, construct a lighting apparatus with instructions from your teacher. This apparatus will probably require the use of fluorescent shop light fixtures suspended above the hydroponic systems. The apparatus should allow you to adjust the vertical height of the lights as the plants grow. See **Fig. 8-14**.

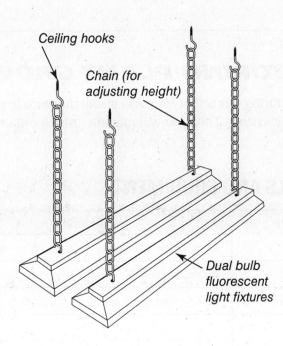

Ceiling hooks

Chain (for adjusting height)

Dual bulb fluorescent light fixtures

Fig. 8-14 Adjustable chains are fastened to ceiling hooks to suspend fluorescent light fixtures over the plants.

3. Be sure that the growing environment is the same for all hydroponic systems so that the principal variables affecting growth will be a result of the systems themselves.

4. Excessive heat, cold drafts, and low humidity all affect plants negatively. Ideally, the indoor relative humidity should be about 40 percent, though this can be difficult to control. Lighting and temperature, on the other hand, may be controlled rather easily with timers and thermostats. Discuss with your teacher how you will be using these devices to control growing conditions.

Project 8 Evaluation

NOTES

THINKING CRITICALLY

1. How do you think commercial hydroponic farms handle the problem of providing enough light?

   _____

   _____

2. Do you think fruit trees could be grown using hydroponics? Why or why not?

   _____

   _____

   _____

MONITORING PLANT GROWTH

Science

During this activity you will install the seeds in your hydroponic system. Then you will monitor growth conditions.

MATERIALS AND EQUIPMENT

Quantity	Description
3.8 L	basic nutrient solution
1	measuring cup
2 packages	quick-germinating seeds, such as radish, bean, or pea
1 roll	litmus paper tape
1 or more	thermometer
1 or more	hygrometer
1 or more	light meter

1. Plant twelve seeds in your hydroponic system.

2. Place your system in the growing area.

3. Obtain your quota of mixed nutrient solution.

4. Most seeds take several days to a week for germination. Be sure they are kept warm and damp during this period. Monitor the humidity, temperature, solution levels, and pH daily. Look for any signs of growth. Be patient, because plant growth is unpredictable. If possible, make your observations at the same time each day. Use the Seed Germination table to keep a record of your observations. If no growth is visible, write "none."

SEED GERMINATION

Day	pH	Humidity	Temperature	Growth/Height
1				
2				
3				
4				
5				
6				
7				
8				
9				
10				

5. Seven to ten days after you have planted your seeds, they will be ready for thinning. Follow the instructions in "Determining Uniform Plant Height," which follows.

THINKING CRITICALLY

1. Did any of the variables you observed change much from day to day? If so, which one? Why do you think this happened?

_____

_____

_____

2. Did all your plants show sprouts on the same day? If not, why do you think this happened?

_____

_____

DETERMINING UNIFORM PLANT HEIGHT

During this activity you will calculate uniform plant height of your seedlings. You will then thin the seedlings, removing those that do not meet the standard.

Project 8 Evaluation

NOTES

MATERIALS AND EQUIPMENT

Quantity	Description
1	metric rule
1	calculator
several	pencils and sheets of paper

1. Some of your seeds may not have sprouted. Discard those. Measure the height of the others. If the plants are anchored in an aggregate, height should be measured from the top of the aggregate to the top of the plant. Be careful not to damage the plants, which are fragile. Record the heights of all the plants in the Seedling Height table.

SEEDLING HEIGHT

Plant	Height
1	
2	
3	
4	
5	
6	
7	
8	
9	
10	
11	
12	
Total	
Mean height	

2. Graph the heights of all the seedlings in your system.

3. Calculate the mean height. Determine a suitable range for uniform plant height:

_____

_____

Project 8 Evaluation

4. Compare your results to those of other teams. What percent of the classes' plants fell within this range? What percent fell outside this range?

_____

_____

_____

_____

5. Thin your seedlings, discarding any that do not fall within the uniform plant height range.

6. Return your hydroponic system to the growing area. After one month the results will be evaluated. In the meantime, resume maintenance and monitoring. Consider the following:
- Are the plants getting enough or too much water?
- Is it time to make fresh solution?
- Are the plants getting enough light?
- If you are using artificial lights, are they working properly?
- Is the growing area free from drafts?
- Is the room temperature—both day and night—at the proper level?
- Is the humidity at the proper level?
- Do the plants seem to be healthy, or do they look spindly and pale?

7. Use separate sheets of paper to create a series of tables like the one shown below for use each day for the next 30 days as you monitor your system.

Plant No.	Today's Date:
Water/solution level	
Amount of light	
Any drafts?	
Temperature	
Humidity	
General health	

Project 8 Evaluation

NOTES

THINKING CRITICALLY

1. What trend(s) appear(s) in the data?

_____

_____

2. Why is it useful to identify a range for uniform plant height in this experiment?

_____

_____

_____

CALCULATING TOTAL PLANT GROWTH

One month has passed. You will now evaluate the effectiveness of your hydroponic system by calculating total plant growth. The team with the highest total growth will win the Challenge.

MATERIALS AND EQUIPMENT

Quantity	Description
1	metric rule
1	calculator
several	pencils and sheets of paper

1. Measure the height of each plant. If the plants are anchored in an aggregate, height should be measured from the top of the aggregate to the top of the plant. Be careful not to damage the plants. Record the heights of all the plants on the Plant Growth Values table.

2. Count the leaves on each plant. Record this number on the Plant Growth Values table.

3. Add the number of leaves to the plant's height to obtain total growth for each plant. For example, a plant that is 10.6 cm tall with eight leaves would have a plant growth value of 18.6.

4. Add all the numbers in the Total Growth column to arrive at the total growth for your entire system.

PLANT GROWTH VALUES

Plant	Height	Number of Leaves	Total Growth
1			
2			
3			
4			
5			
6			
		Total for System	

5. Was the combination of plant height and number of leaves an appropriate measure? Why or why not?

_____

_____

_____

THINKING CRITICALLY

1. How would you design another hydroponic system differently?

_____

2. What data other than plant height and number of leaves might have been collected to help determine the winner of the Challenge?

_____

ANALYZING DATA

Math

During this activity, you will analyze the data you have been collecting throughout the project. You will prepare charts and graphs to present the data visually.

MATERIALS AND EQUIPMENT

Quantity	Description
assorted	sample charts and graphs
1	calculator
several sheets	graph paper, 1 cm or ¼″

Project 8 Evaluation

NOTES

1. As a class, brainstorm and discuss various means of visually representing data collected during this project. Look at sample charts and graphs for ideas.

2. Discuss the advantages and disadvantages of the various methods.

3. Prepare the kinds of tables, charts, graphs or other means you think are useful and appropriate for communicating your team's results from the Challenge.

4. For each visual representation your team creates, write a concise, two- or three-sentence summary statement. Teammates may discuss their ideas before arriving at their final summary statements.

5. Prepare the following graphs to represent your team's progress in the Challenge. Then, as a class, discuss the graphs.
   • Time compared to overall plant growth, with time on the horizontal axis. See **Fig. 8-15**.
   • Time compared to the number of leaves grown, with time on the horizontal axis.
   • Time compared to root growth, with time on the horizontal axis.
   • Time compared to stem and leaf growth (above the "ground"), with time on the horizontal axis.
   • Stem and leaf growth compared to root growth, with stem and leaf growth on the horizontal axis.

6. Compute the mean growth for the plants in your hydroponic system as well as the class mean. How does your mean compare to the class mean?

_____

7. Compute the mean number of leaves per plant for your hydroponic system as well as the class mean. How does your mean compare to the class mean?

_____

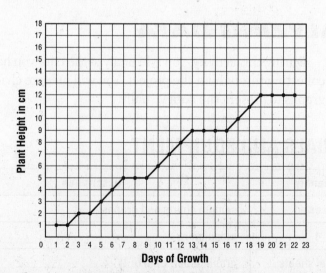

Fig. 8-15 A sample graph

Project 8 Evaluation

THINKING CRITICALLY

1. How does the design of the various systems seem to be related to plant growth? That is, are there design features that seem to lead to increased plant growth? (Use your data to draw your conclusions.)

_____

2. As data are collected, how does this evidence compare with your original hypotheses regarding plant growth?

_____

_____

ANALYZING COSTS

For this activity, you will analyze the cost of your hydroponic system. You will then determine what makes similar undertakings cost effective. Actual costs fall into three major categories:

MATERIALS AND EQUIPMENT

Quantity	Description
1	paper and pencil
1	calculator

- Direct materials cost—the cost of the materials that actually end up in the hydroponic system and the cost of supplies used, such as plastic, sheet metal, fasteners, seeds, and nutrient solution
- Direct labor costs—the cost of labor for those people needed to plant, maintain, and harvest the crop
- Overhead—includes facility costs, such as rent and utilities, as well as the costs of indirect materials and indirect labor (such as supplies, abrasive paper and saw blades, assistants and sales force.)

1. As a class, brainstorm a list of direct materials and utility costs.

2. Determine the direct materials cost for constructing your team's hydroponic system. If the material was recycled or you do not know the cost, use your best guess. Write the information on the Direct Materials Cost table. If you need more room, use a separate sheet of paper.

DIRECT MATERIALS COSTS

Material	Unit Price	Total
Total Cost		

Project 8 Evaluation

NOTES

3. Determine the costs of direct materials for making 1,000 hydroponic systems like yours. Assume that you will receive a 10 percent discount because you are purchasing large quantities and that there will be a 5 percent waste of materials.

4. Even if you did not actually have to use artificial light, assume you used four 40-watt dual bulb fluorescent fixtures as grow lights for your system. Call your local water and electric companies to determine local costs per unit (gallon and kilowatt-hour, respectively). Then, given the local cost of electricity (per kilowatt-hour) and the fact that each 40-watt fluorescent bulb draws 40 watts of electricity every hour it is on, estimate the amount of electricity that your hydroponic system would require for a month of operation.

5. Estimate the monthly cost of lighting 1,000 hydroponic systems like yours.

6. How do you think the costs of growing plants hydroponically compare with the costs of growing a similar number of plants with conventional (soil-based) agricultural techniques?

_____

_____

_____

THINKING CRITICALLY

1. How do you think the costs/profits of growing plants with 1,000 hydroponic systems like the one you designed might compare to the costs/profits of growing plants with 10,000 hydroponic systems?

_____

_____

_____

_____

2. Why should you maintain skepticism when reading cost estimates of projects, social programs, and other projects in the newspaper?

_____

_____

_____

Digital Photography

Design Brief

The word photography means "writing with light." Light is used to capture images with a camera so that the images can be reproduced. (See **Fig. 9-1**.) Digital photography combines traditional photography with the advantages of a computer. Photos can be downloaded into the computer and altered in many ways before they are printed.

Any photograph can provide a large amount of information about a tiny segment of time and space. However, when used together in a photo essay, several photographs can tell us even more. They can communicate ideas.

Fig. 9-1 Photographic images can provide information and tell stories.

Challenge

Design and create an effective photo essay that informs, entertains, or persuades your audience about a topic of your choice.

Criteria

- Your photo essay should consist of a series of five to ten photographs with captions and title.
- The photographs should be mounted and displayed in a manner that supports the theme of the photo essay.
- You must keep a record of each of the photographic techniques you decide to use and why you think they are effective.
- All projects must be accompanied by the appropriate documentation, including the following:
    - ☐ Sketches or storyboard of all possible design solutions considered
    - ☐ A list of the topics you considered for the photo essay along with a short explanation of each topic
    - ☐ An explanation of how and why you chose your particular topic

NOTES

☐ A short description of each image's significance to the topic
☐ Exposure data for each of your images, including: time, location, shutter speed, f-stop, and ISO value
☐ Information gathered from resources
☐ Any notes made along the way

Constraints

- The content of your essay must be appropriate for a school audience. Before beginning, check your subject with your teacher.
- Your finished product must not exceed the size of the display space.
- Images should not be too small or too dark to be easily viewable.

Engineering Design Process

1. **Define the problem.** Write a statement that describes the problem you are going to solve. The Design Brief and Challenge provide information.
2. **Brainstorm, research, and generate ideas.** The Research & Design section, beginning on page 275, provides background information and activities that will help with your research.
3. **Identify criteria and specify constraints.** Refer to the Criteria and Constraints listed on pages 273 and 274. Add any others that your teacher recommends.
4. **Develop and propose designs and choose among alternative solutions.** Remember to save all the proposed designs so that you can turn them in later.
5. **Implement the proposed solution.** Once you have chosen a solution, determine the processes you will use to make your photo essay. Gather the tools and materials you will need. Make sure you understand and follow all safety rules.
6. **Make a model or prototype.** The Modeling section, beginning on page 288, provides instructions for completing the essay.
7. **Evaluate the solution and its consequences.** The Evaluation section, beginning on page 290, describes how to evaluate your essay.
8. **Refine the design.** If instructed by your teacher, make changes to improve your design.
9. **Create the final design.** If instructed by your teacher, improve the existing essay or create a new one based on the revised design.
10. **Communicate the processes and results.** Write a report about your project. Be sure to include all documentation listed under Criteria.

Research & Design

In this section you will learn about light and cameras and how they can be used to create interesting and effective photographs. This information can help you prepare possible essay designs and then decide which design is most likely to meet the criteria and constraints.

LIGHT, LENSES, AND FILTERS

Light is a form of energy that radiates outward in all directions. Some things, such as a light bulb, create their own light. Most things, however, only reflect light that originates elsewhere. You can see a book you are reading because light rays bounce off the page and reach your eyes. When you use a camera, the light rays bounce off the subject and reach the camera lens. A **lens** is a piece of curved and polished transparent material, such as glass, that refracts (bends) and focuses light.

Lenses may be concave (curved inward), convex (curved outward), or compound (a combination of both). See **Fig. 9-2**. Most modern cameras use compound lenses.

MATERIALS AND EQUIPMENT

Quantity	Description
1	flashlight
several	objects for viewing through lenses
1	prism
several	examples of lenses, both simple and compound
1	camera
assorted	filters for color photography

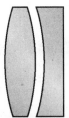

Convex   Concave   Compound

Fig. 9-2 Lenses come in many shapes. A lens may have one flat surface and one concave or convex surface or both.

Name_____ Date_____ Class _____

NOTES

Lens	Type of Lens	Observations
A		
B		
C		
D		

1. View different objects through various lenses. Then, note your observations.

2. Darken the room. Turn on the flashlight and place it about 5 cm behind each lens so it shines through them. Note how each lens affects the light.

3. Darken the room. Shine the flashlight through the prism. What happens to the light? Why?

_____

Filters are used for photography for several reasons. They may be used to match the color balance of the light so a scene looks more realistic. For example, a blue sky may appear very light in a black and white photo. Using a deep yellow filter can darken the sky and make any clouds stand out. A fluorescent filter can decrease the greenish cast of color pictures taken under fluorescent light.

4. Attach a filter to the camera and, using the appropriate settings, take a picture. If you are unfamiliar with camera settings, ask your teacher for a demonstration.

5. Repeat with all the examples of filters you have been given. Note which filters you used on the table.

6. Print the photographs. Note the effects created by the filters.

Photo No.	Type of Filter	Effects

THINKING CRITICALLY

1. In general, what is it that the filters are filtering out?

_____

2. What effect does a concave lens have on light rays? A convex lens?

_____

ADJUSTING THE CAMERA

NOTES

Engineering Technology

Modern cameras share the same set of basic parts: a lens, an aperture, a shutter, a light-tight box, and a recording medium.
- The lens refracts light and focuses the image on the recording medium.
- The aperture controls the amount of light that reaches the recording medium. The aperture can vary in size. The larger the opening, the more light can enter the camera.
- The shutter controls the amount of time that light is allowed to reach the recording medium.
- The light-tight box ensures that only the light from the lens reaches the recording medium.
- The recording medium captures the image. In a digital camera, the recording medium is a sensor. In a film camera, it is the film.

Many cameras also include an integrated light meter, a flash unit or a hot-shoe on which an external flash can be mounted, and a central processing unit that makes decisions about focus and exposure based on the information received from sensors.

Refer to your camera's instruction manual to learn how to use a built-in light meter and make other settings to determine proper exposure. When your image is in focus and your camera's meter indicates you have selected the proper shutter speed and aperture combination, you are ready to take the picture.

MATERIALS AND EQUIPMENT

Quantity	Description
1	camera with a variable focal length or interchangeable lenses and manual and automatic setting capabilities
1	tripod
1	light-colored subject
1	dark-colored subject
1	light-colored background
1	dark-colored background
1	computer and printer

SAFETY FIRST

Before You Begin Make sure you understand how to handle the camera and equipment without damaging them. Ask your teacher to demonstrate their proper use. Follow all rules for your own safety. As you are looking into the camera, be aware of where you're standing or walking.

NOTES

The Lens

After the light rays pass through the camera's lens, they come together inside the camera at the **focal point**. The distance (in millimeters) from the center of the lens to the focal point, located on the recording medium, is the **focal length**. See **Fig. 9-3**.

The longer the focal length of the lens is, the narrower its angle of view is (viewing range) and the greater the degree of magnification. A "normal" lens has an angle of view approximately equal to that of the human eye (46°). Lenses with an angle of view that is wider than normal are referred to as wide-angle lenses. Lenses with an angle of view that is narrower than normal are considered telephoto (magnifying) lenses.

Lenses are also classified according to the size of the aperture used. Divide the focal length by the lens's diameter to find maximum aperture.

1. Insert a lens with a "normal" focal length into the camera.

2. Place the camera on a tripod and put it in manual exposure mode. Adjust aperture and shutter speed for proper exposure. Take a picture of a stationary subject.

3. Insert a lens with a short focal length into the camera. Photograph the same stationary subject.

4. Without moving the camera, photograph the same subject again using a lens with a long focal length.

5. Print the images and describe the results:

_____

_____

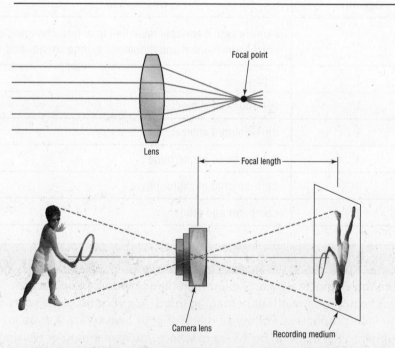

Fig. 9-3 The focal point is where the light rays meet. The focal length is the distance inside the camera from the lens to the recording medium.

The Light Meter

Taking the recording medium's sensitivity to light into account, the light meter suggests a shutter speed and aperture setting. Many cameras will set them automatically unless instructed otherwise. Are these electronic decisions always correct? Not always, of course. The "right" settings are determined by the effect that the photographer is trying to achieve.

1. Place the camera on a tripod and put the camera in automatic exposure mode.

2. Place a light-colored subject in front of a light-colored background and take a picture.

3. Place a light-colored subject in front of a dark-colored background.

4. With no change in camera settings or lighting, take a picture.

5. Place a dark subject in front of a dark background and with no change in camera settings or lighting, take a picture.

6. Print the photos.

THINKING CRITICALLY

1. Does the light meter record the light falling upon the subject or the light being reflected from the subject into the camera?

_____

_____

2. Did the photographs have the same exposure? Why or why not?

_____

_____

_____

The Aperture

Math

The aperture is an opening in the lens, either in front of it or within the lens elements. Changing the diameter of the aperture gives the photographer a means of controlling the amount of light that reaches the recording medium.

Aperture settings are indicated by a series of numbers called **f-stops**. The bigger the number, the smaller the opening. An aperture value of f/1.4 will let more light pass through the lens than an aperture value of f/2.8. See **Fig. 9-4** on page 280.

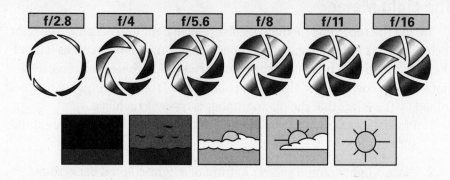

Fig. 9-4 The aperture increases or decreases the amount of light allowed into the camera. Outdoor conditions affect aperture settings.

If the camera's light meter indicates that you need more light for proper exposure, you can widen the aperture. On the other hand, if the camera's meter indicates that you need less light for proper exposure, you can narrow the aperture. Suppose your aperture is set at f/2.8 and you need more light. Which setting would you use, f/1.4 or f/4.0?

_____

1. Place the camera on a tripod and point it at a stationary subject.

2. Put the camera in manual exposure mode. Adjust aperture and shutter speed for proper exposure. Take a picture.

3. Now, change the aperture to one f-stop higher than the current value. Do not change the shutter speed. Take a picture.

4. Raise the f-stop again. Do not change the shutter speed. Take a picture.

5. Return the aperture to the value at which it was set in Step 2, then set the aperture one f-stop lower than the current value. Do not change the shutter speed. Take a picture.

6. Lower the f-stop again. Take a picture.

7. Print the images, compare the results, and record your observations:

f-stop Setting	Appearance of Image

The aperture also plays a part in depth-of-field. **Depth-of-field** refers to the area of the image, from near to far, that is in focus. Three factors determine depth-of-field: the focal length of the lens, the f-stop setting, and the camera-to-subject distance.

If all other things remain equal, as the focal length (distance from the lens to the recording medium) decreases, the depth-of-field increases. A 135 mm telephoto lens has a much shallower depth-of-field than a standard 50 mm lens.

8. Place the camera on a tripod about 15 feet from the subject.

9. Place masking tape or some other markers at 3-foot intervals for a distance of 30 feet or more from the camera outward, passing beyond the subject. Write the distance on each piece of tape.

10. Adjust the lighting, f-stop, and shutter speed.

11. Using the same subject, f-stop, shutter speed, and lighting conditions, take pictures using two or more different lenses having different focal lengths. (A zoom lens will also work.)

12. Print the images. Estimate and graph the range of distances for each lens in which there is sharp focus, such as from 3 to 6 feet. See **Fig. 9-5**.

Lens	3	6	9	12	15	18	21	24	27	30
A										
B										
C										
D										

Fig. 9-5 One possible way to graph depth-of-field data

Using an aperture with a small f-stop value will generally produce an image with a very shallow depth-of-field. Using an aperture with a large f-stop value will generally yield an image with a very deep depth-of-field. Suppose you want the shallowest depth of field. Which setting would you use, f/2.8 or f/1.4?

_____

13. Place a 50 mm lens in the camera. Adjust the lighting, f-stop, and shutter speed for proper exposure. Take a picture.

14. Using the same subject, lens, speed, and lighting conditions, take pictures using five or more different f-stops.

15. Print the images. Estimate and graph the range of distances for each f-stop setting in which there is sharp focus. What other effects were noticeable in the printed photos as the aperture got larger?

_____

Name_____ Date_____ Class _____

NOTES

Depth-of-Field Data		

Focal Length (in mm)

Distance within Focus (in feet)

16. Next, change the camera-to-subject distance. Using the same subject, lens, speed, f-stop, and lighting conditions, move the subject to five different distances from the camera. Take the pictures.

17. Print the images. Estimate and graph the range of distances for each position in which there is sharp focus.

Shutter Speed

The length of time the shutter is open to allow light to reach the recording medium is referred to as **shutter speed**. Speeds may vary from as slow as 30 seconds to as fast as 1/8000 of a second. The most common range of speeds is from 1 second to 1/1000 of a second. The smaller the fraction, the shorter the amount of time the shutter is open. Each setting is either half or twice as long as the one next to it.

Shutter speed and f-stop settings work together. In low light, you might want a wide aperture and a slow shutter speed so that as much light as possible enters the camera. Different combinations create different effects.

1. Place the camera on a tripod and point it at a stationary subject.

2. Put the camera in manual exposure mode, meter the scene, and set the appropriate combination of aperture and shutter speed for proper exposure. Take a picture.

3. Now, change the shutter speed to one step faster than the current value. Do not change the aperture. Take a picture.

4. Increase the shutter speed by one step again. Take a picture.

5. Reset the shutter speed to the value in Step 2.

6. Now, change the shutter speed to one setting slower than the current value. Do not change the aperture. Take a picture.

7. Decrease the shutter speed by one step again. Take a picture.

8. Print the images, compare results, and record observations:

Shutter Speed	Appearance of Image

ISO Values

The sensitivity to light of the recording medium is based upon the standards developed by the International Organization for Standardization (ISO). For film cameras, this value refers to film "speed." In a digital camera, it refers to a sensor adjustment. The range of speeds for a fairly simple camera is usually between ISO 50 and 800.

A recording medium with an ISO value of 100 is twice as sensitive to light as one with a value of 50 and half as sensitive as one with a value of 200. Bright conditions usually require a low ISO number. Darker conditions require higher numbers.

1. Place the camera on a tripod and point it at a stationary subject.
2. Put the camera in manual exposure mode, meter the scene, and set the appropriate combination of aperture and shutter speed for proper exposure. Take a picture.
3. Now, change the shutter speed to one step faster than the current value. Do not change the aperture. Take a picture.
4. Increase the shutter speed by one step again. Take a picture.
5. Reset the shutter speed to the value in Step 2.
6. Now, change the shutter speed to one setting slower than the current value. Do not change the aperture. Take a picture.
7. Decrease the shutter speed by one step again. Take a picture.
8. Print the images, compare results, and record observations on the table.

Shutter Speed	Appearance of Image

THINKING CRITICALLY

1. If the amount of light and the ISO value of the recording medium are constant, what are the variables for an exposure?

_____

2. The following settings will all result in the same exposure. Fill in the missing numbers.

f-stops	Shutter speeds
f/11	1/30 second
f/8	1/60 second
	1/125 second
f/4	
	1/500 second

NOTES

PHOTOGRAPHIC TECHNIQUES

Taking interesting pictures means more than just pointing and clicking. Various accessories and techniques allow you to control the process.

MATERIALS AND EQUIPMENT

Quantity	Description
1	tripod
1	camera with built-in or flash attachment and both automatic and manual setting capabilities

Flash Photography

A built-in flash or flash attachment provides extra light. Modern flashes often contain complex electronics that measure the light reflected from the scene and communicate with the camera. The flash itself lasts for only a fraction of a second, just long enough for a proper exposure. The flash exposure can also be controlled by varying the camera's aperture. Small apertures let in less light and large apertures let in more light.

When the flash is discharged, light travels from the flash to the subject and is reflected back to the camera's lens. The amount of light that reaches the subject from the flash is inversely proportional to the square of the distance from the light source to the subject. It is written as $1/r^2$ (r represents the distance from the flash to the subject). If the distance is 8 feet, the amount of light would be $\frac{1}{8} \times 8$, or $\frac{1}{64}$ the intensity of the light at 1 foot.

Using the equation $1/r^2$, if you double the distance of the flash to the subject, the amount of light reaching the camera decreases to one quarter that of the original distance. See **Fig. 9-6**. Light at the source is measured in foot-candles (illumination of one candle measured from one foot away).

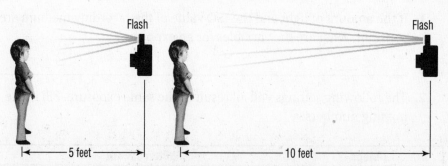

Fig. 9-6 Because light rays scatter, the amount of light reaching the subject decreases.

1. What is the brightness of a 4,000-foot-candle light source that is one foot away?

_____

2. If a light source measures 1,000 foot-candles from 2 feet away, how bright might it be at a distance of 10 feet?

_____

_____

3. Using the formula, calculate the number of foot-candles you would expect at 1.75 feet

_____

_____

STOP-ACTION PHOTOS

Math

Have you ever seen a photo of a runner "frozen" in mid-step? This is called stop-action photography. See **Fig. 9-7**. A fast shutter speed can freeze motion. A slow shutter speed allows the image to blur. How this technique is used depends on what you want to show. Sometimes freezing motion can be interesting if a runner, for example, seems to be leaping in the air. Panning, moving the camera along with the subject, can also keep movement in sharper focus. At other times, allowing the motion to blur can increase the impression of speed.

1. Place the camera on a tripod. Set the shutter speed at 1/30 sec, or another appropriately slow setting. Have someone walk fast past the camera and take a picture.

2. Set the shutter speed at a faster setting, such as 1/125 sec and take a similar picture.

3. Remove the tripod. Reset the shutter at 1/30 sec and pan along with the person as someone walks fast past you. Take the picture.

4. Print the pictures and compare the results.

Fig. 9-7 Some photo techniques can freeze motion.

NOTES

THINKING CRITICALLY

1. At what setting did the action appear frozen?

_____

2. At what setting did a blur suggest speed?

_____

COMPOSITION

Composition is the way in which the parts of an image are arranged. For example, a photo of mountain scenery might be composed with the mountains in the background and a twisted tree in the foreground. Composition determines what is emphasized and whether or not the arrangement is pleasing to the eye.

One technique of composition is known as the rule of thirds. The area of the photograph is divided into thirds both horizontally and vertically. See **Fig. 9-8**. By positioning your subject near one of the spots where the lines intersect, you can create a more interesting composition.

Framing the subject in a photograph can also lend visual appeal. Try using elements in your subject's environment. For example, a photo of a sunset might be framed by tall palm trees on either side. Also, consider where your subject is in relation to the edges of the image.

Pay attention to what is in the background and the foreground of the picture. Make sure that nothing distracts from your subject.

Using leading lines can draw the viewer's eye to the subject of your photograph. If there's a roadway in your photo, for example, your photo might show the road leading into the distance toward a city.

Fig. 9-8 The subject is located at one of the points where lines intersect.

Name_____ Date_____ Class _____

THINKING CRITICALLY

1. If you were using the rule of thirds to compose a person's portrait, which facial feature would you place at the intersection of lines?

_____

2. Why would it be reasonable to assume that not all of the source light is actually shining on a photographer's subject?

_____

DESIGNING YOUR PHOTO ESSAY

Engineering Technology

Now that you have learned how to take effective photos, you can design your photo essay. Remember that you can inform, entertain, or persuade your audience about your topic. For example, suppose you ride a bike to school and think the school should have more bike parking. You could inform your audience that a problem exists. You could entertain your audience with humorous ways in which students have solved the problem by themselves. You could attempt to persuade school officials that more parking is needed.

1. Do research to decide what is most important to photograph.

2. Sketch storyboards for several possible essays on your topic. See **Fig. 9-9**. Be sure to include several techniques you have learned about, such as changing depth of field or stop-action shots.

3. Select the storyboard that you think is the most effective.

4. Check your storyboard against the Criteria and Constraints.

Fig. 9-9 A storyboard shows your photo ideas in the proper arrangement.

THINKING CRITICALLY

1. What special techniques will you use when taking your photos?

_____

2. Which approach will you be taking for your topic: informing, entertaining, or persuading?

_____

NOTES

Modeling

During this part of the project, you will create your photo essay. When the images for your photo essay have been chosen and printed, you will need to mount and display them.

SAFETY FIRST

Before You Begin Make sure you understand how to handle the camera and other equipment without damaging them. Ask your teacher to demonstrate their proper use. Follow all rules for your own safety as well. As you look through the camera, watch where you are standing or walking. If you want to photograph people, get their permission first.

CREATING YOUR PHOTO ESSAY

Mounting your prints makes them look professional. It also makes it easier to display them. Dry mounting uses a special type of adhesive to permanently attach a print to a medium such as matte board. The print and a sheet of adhesive are cut to size, positioned on the matte board, and placed in a press. Heat activates the adhesive, and pressure forces the adhesive into the paper and matte board.

Prints can also be mounted using photo corners. Photo corners are small adhesive pockets placed on the mounting medium to hold the print in place by securing its corners. Prints mounted with photo corners can usually be removed from the mounting medium if necessary.

Once the prints are mounted, they can be over-matted. A second sheet of matte board with a beveled window, cut to match the size of the print, is placed over the top of the print and secured. See **Fig. 9-10**.

Prints should be arranged in a way that suits the topic of your photo essay. How they are arranged can also lead the viewer's eye from one photograph to another. For example, if you were to show the day-by-day construction of a building, you might want to arrange your images in chronological order. If you wanted to campaign against litter in your school, start with minor examples and end with the worst.

1. Take your photos and print them.

2. Mount your photos using your chosen method.

3. Write captions for the photos and create any other needed graphics.

4. Arrange the photos and captions in the way that best tells your story.

5. Give your essay a title.

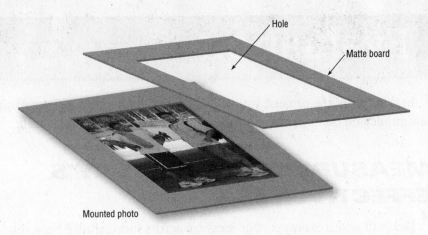

Hole

Matte board

Mounted photo

Fig. 9-10 A second sheet or matte board is placed over the photo like a frame.

Quantity	Description
1	camera and accessories, as needed
1	computer and photo-quality printer
assorted	mounting media
assorted	display materials, as needed

THINKING CRITICALLY

1. In what other ways could a photo essay be mounted and displayed?

_____

_____

_____

2. In what order would you display photos to build support for the basketball team?

_____

_____

_____

NOTES

Evaluation

In this part of the project, you will determine how well the photo essays communicated their message. You will then discuss the methods used.

MEASURING YOUR ESSAY'S EFFECTIVENESS

Did your photo essay get your message across successfully? Now you will find out how your audience responds.

1. Your teacher will assign you the photo essay created by another student. Review that student's essay by answering the questions and completing the table. In the ratings table, 1 is a low score; 5 is high.
   - What is the title of the essay?
   - Was the purpose to inform, to entertain, or persuade?
   - Was the purpose achieved? Explain.
   - What was the photographer trying to say about the topic?
   - Which photo in the essay caught your eye first? Why?

	1	2	3	4	5
Special techniques used effectively					
Shows understanding of camera settings					
Captions and title are clear and effective					
Photos are well mounted					
Display is neat and well thought out					

2. As a class, discuss the photo essays, the methods used, and audience responses. Be courteous when discussing essays done by others.

THINKING CRITICALLY

1. Which part of this project did you find the easiest? Which part of this project did you find the hardest? Explain why.

_____

_____

2. If you were to repeat this project, what would you do differently?

_____

_____

General Safety Rules

In a technology course, you have many opportunities to design and build products. You can apply your creativity and problem-solving skills. It is very important that you also apply common sense and practice safe work habits.

Develop a Safe Attitude

- Read and follow all posted safety rules.
- Take the time to do the job right.
- Consider each person's safety to be your responsibility. Avoid putting others in danger.
- Stay alert. Be aware of your surroundings.
- Work quietly and give your full attention to the task at hand. Never indulge in horseplay or other foolish actions.
- Stay out of danger zones as much as possible. These are usually marked with black and yellow striped tape.

- Put up warning signs on things that are hot and could cause burns.
- If you bend down to pick up an object, use your legs, not your back, to lift up. Keep your back straight. To keep better control, get help to lift or move long or heavy items.
- Handle materials with sharp edges and pointed objects carefully.
- Report accidents to your teacher at once.

» Notice the safety jackets worn by the students working on this solar-powered car.

Have Respect for Tools and Equipment

- Never use any tool or machine until the teacher has shown you how to use it and has checked the setup.

- Before using any tool or machine, make sure you know the safety rules and make sure you get your teacher's permission.

- Use equipment only when the teacher is in the lab.

- Do not let others distract you while working.

- Do not use electrical tools or equipment if the cord or plug is damaged.

- Always use the right tool for the job. The wrong tool could injure you or damage the part you are working on.

- To avoid injury, use the right machine guard for the job. Check with your teacher for the appropriate guard.

- Keep hands and fingers away from all moving parts.

- Before you leave a machine, turn it off and wait until it stops. If you are finished, clean the machine and the area around it.

- When you have finished working, return all tools and any unused supplies to their proper places.

Prevent and Control Fires

- Store oily rags in a closed metal container to prevent fire.

- Know where the nearest fire extinguisher is and how to use it, *if that is your school's policy.*

Wear Appropriate Clothing and Protective Equipment

- Always wear eye protection. Special eye protection may be needed for some activities, such as using a laser, welding, or using chemicals.

- Wear hard shoes or boots with rubber soles.

- Use ear protection near loud equipment.

- Do not wear loose clothing, jewelry, or other items that could get caught in machinery. Tie back long hair.

- Do not wear gloves while operating power tools.

Have Respect for Hazardous Materials and Waste

- Products with major health risks should have a Material Safety Data Sheet (MSDS) available. Ask your teacher about the MSDS before you use materials that may be hazardous. Know how to read the MSDS. You can find more information about Material Safety Data Sheets on page 7.
- Other common chemicals will have safety information on the labels. Check the labels of materials before you use them.
- Wear appropriate personal protective equipment (PPE) when working with hazardous materials.

Safe Use of Hand Tools

Misuse and improper maintenance pose the greatest hazards in using hand tools. Observe the following rules when using hand tools.

- Keep all hand tools clean and in good condition.
- Wipe tools clean before and after use.
- Use only tools that are in good condition. If a tool is damaged, tell your teacher about it.
- Use a hand tool only for the purpose for which it was designed.
- Hold and use the tool in the proper way, following manufacturer's instructions.
- Always wear safety glasses or goggles.
- Cutting tools should be sharp.

- Work in a well-ventilated area.
- Follow your teacher's instructions for disposal of hazardous materials and waste.

Maintain the Lab

- Keep the work area clean. Keep the floor and aisles clean at all times.
- If a liquid is spilled, clean it up immediately as instructed by the teacher.
- Always use a brush, not your hands, to clean dry materials from a table or piece of equipment.
- Store all materials properly.

- Do not use a screwdriver on a part that is being hand-held. The screwdriver can slip and hit your hand.
- Carry sharp or pointed tools with the point down and away from your body.

>> This carpenter is nailing roof trusses. *Why should eye protection be worn when using a hammer?*

Safe Use of Electric Power Tools

Use care and common sense when working with electric power tools. Observe the following safety practices.

Prevent Electric Shock

- Do **not** stand in water while working on equipment.
- Make sure that all electrical cords are free of frays and breaks in the insulation.
- Pull the plug, not the cord, when you unplug a tool or machine. Damaging a cord may cause an electric shock.
- Keep electrical cords away from sharp edges.
- Use power tools that are properly grounded or double insulated.
- Make sure that all extension cords are the three-wire grounded type.
- Make sure the three-pronged plug is used in a grounded receptacle.

Work Safely

- Do **not** wear loose clothing, ties, or jewelry that can become caught in moving parts of machinery. Be sure to tie back long hair.
- Do **not** wear gloves while operating power tools.
- Wear the appropriate personal protective equipment (PPE), such as safety glasses or a face shield.

- Never set a hand-held power tool down while it is running or coasting.
- Avoid accidental startups by keeping fingers off the START switch when carrying a tool.
- When you approach a machine, be sure it is off and that it is not coasting.
- Secure the work piece with clamps or a vise. This will free both hands to operate the tool.
- Disconnect power source before changing accessories such as bits, belts, and blades.
- Keep tools as sharp and clean as possible for best performance.
- Tell the teacher immediately if the machine does not sound right or if you can see that something is wrong.

Safe Use of Cutting Tools

Observe the following safety practices when using electric power tools to cut materials.

- Allow machines to reach full speed before starting to cut.
- Before working on stock (wood or other workpieces), check it for cracks, loose knots, and nails.
- The shortest piece of lumber that can run safely through most equipment is 12 inches long.
- Keep your balance; don't overreach.
- Support ends of long stock before cutting.
- Wait until the blade stops completely before removing scraps. Use a brush, not your fingers.

Safe Use of Pneumatic Tools

Pneumatic tools are powered by compressed air. The air is fed to the tool through a high-pressure hose connected to an air compressor.

- When using a compressed-air gun, always wear the correct personal protective equipment (PPE). For example, wear safety glasses or goggles and a face shield. Always direct the airflow away from you.
- Always carry a pneumatic tool by its frame or handle, not by the air hose.
- Make sure all pneumatic tools are securely attached to the compressed-air line.

- Never use a pneumatic tool, such as a compressed-air gun, to remove debris from your clothing or body.
- Securely position a pneumatic tool before operating it. These tools operate at high speeds or under high pressure.

Tool and Machine Maintenance

Well-maintained tools and machines do a better, faster job and are safe.

Hand Tool Maintenance

- Keep sharp tools properly sharpened.
- Check edges, sharp points, and other working surfaces for cracks and other defects.
- Keep working surfaces and handles clean and free of dirt and rust.
- Check that any moving parts or mechanisms are working properly.
- Check for loose handles and other loose parts.
- Check wood handles for splinters and damage.
- Clean the tool following use each day.
- Inspect the tool for damage at least once each day before use.
- Lubricate any tool parts that require it.
- Preserve wood handles regularly with special oils made for that purpose.
- Store tools in their proper places in an organized way.
- Store sharp tools in a safe place with the cutting edge protected.

Power Tool and Machine Maintenance

- Keep sharp tools properly sharpened.
- Check edges, sharp points, hoses, and other working parts for cracks and other defects.
- Keep working surfaces, handles, and hoses clean and free of dirt and rust.
- Check that any moving parts or mechanisms are working properly.
- Check for loose handles and loose parts.
- Check for broken plugs or lugs removed from grounding plugs.
- Check for split insulation or damaged electrical cords.
- Disconnect and clean following use each day.
- Disconnect and inspect tools before use.
- Lubricate any tool parts that require it.
- Store and organizetools in their proper places.
- Store sharp tools in a safe place with the cutting edge protected.

Safe Use of Chemicals

Follow these rules when working with paints, stains, varnishes, paint thinners, adhesives, or other chemicals.

- Read and follow all label precautions.
- Always wear protective clothing and approved eye protection. Use appropriate gloves or tongs when needed.
- Many chemicals produce harmful fumes. Work only in well-ventilated areas. Use a respirator whenever it is required.
- Know where the eyewash station is and how to use it.
- Know where the Material Safety Data Sheets (MSDS) are located.
- Know where the poison-control phone number is located.
- Mix chemicals only as directed. If you need to mix acid and water, get the water first. Then carefully add the acid to it.

- Pay attention to others around you when working with chemicals and report any unusual reactions.
- Clean up spills immediately.
- Clean all tools and equipment properly after using.
- Avoid skin contact with chemicals. Wash thoroughly before leaving the area.
- Many chemicals need to be stored away from heat or away from moisture. Follow label directions.
- Never store chemicals in an unlabeled or incorrectly labeled container.
- Store chemical-soaked rags in an approved container.
- Dispose of chemicals properly.

» Use gloves to protect your hands from contact with chemicals.

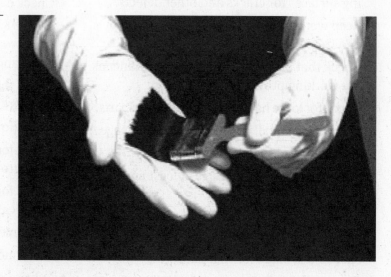

Material Safety Data Sheet

OSHA requires that workers be informed about any hazardous chemicals to which they may be exposed. A Material Safety Data Sheet (MSDS) is a form used to communicate information about hazards. The table at the bottom of this page is part of an MSDS for acetylene. There are different types of Material Safety Data Sheets, but they all must include the following kinds of information.

- The **identity** of the hazardous chemical, including both its chemical and common name(s). If the chemical is a mixture, the ingredients are listed.
- **Physical and chemical characteristics**, such as flash point (the lowest temperature at which a vapor may ignite).
- **Physical hazards**, including the potential for fire, explosion, and reactivity.
- **Health hazards**, including signs and symptoms of exposure, and any medical conditions that are generally recognized as being aggravated by exposure to the chemical.
- The **primary route(s) of entry**. For example, a chemical may produce fumes that could be inhaled.

- The **limit of safe exposure**.
- Whether the chemical is **carcinogenic** (cancer-producing).
- Generally applicable **precautions for safe handling and use**.
- Generally applicable **control measures**, such as personal protective equipment (PPE).
- **Emergency and first-aid** procedures.
- The **date** the MSDS was prepared or last updated.
- The contact information of the chemical manufacturer, importer, employer, or other responsible party who can serve as a **provider of additional information** on the hazardous chemical and appropriate emergency procedures, if necessary.

2. Hazard Ingredients and Identity Information

COMPONENT	% VOLUME	OSHA-PEL	ACGIH-TLV	CAS NUMBER
Acetylene	95.0 to 99.6	Not Available	Simple Asphyxiant	000074-86-2
Acetone	Unavailable	1000 ppm TWA	750 ppm TWA 1000 ppm STEL	000067-64-1

3. Physical and Chemical Characteristics

Boiling Point:	-118.8°F	**Specific Gravity:**	0.906	**pH:**	N/A
Melting Point:	-116°F	**Evaporation Rate:**	N/A	**Physical State:**	Gas
Vapor Pressure:	635 psig	**Solubility (H²0):**	Soluble		

Appearance and Odor:
Pure acetylene is a colorless gas with an ethereal odor. Commercial (carbide) acetylene has a distinctive garlic-like odor.

How to Detect This Substance: N/A

Other Physical and Chemical Data:
Liquid density at boiling point, 38.8 lb/ft³ (622 kg/m³)
Gas density at 70°F 1 atm, 0.0691 lb/ft³ (1.107 kg/m³)

Safety Color Codes

Safety signs and labels are color-coded to signify hazards or to identify the location of safety-related equipment.

Color	Meaning	Examples
Red	Danger, stop, or emergency	• Fire-protection equipment • Flammable-liquid container • Emergency stop bars and switches
Orange	Be on guard	• Hazardous parts of equipment or machines that might injure • Safety starter buttons on equipment or machines
Yellow	Caution	• Physical hazards, such as steps and low beams • Waste containers for combustible materials
White	Storage	• Housekeeping equipment
Green	First aid	• Location of safety equipment, such as first-aid kit
Blue	Information or caution	• Out-of-order signs on equipment • Cautions against using out-of-order equipment

Warning Labels

The warning labels on containers of hazardous material may use colors and numbers to indicate the hazard level. There are several labeling systems. One was developed by the National Fire Protection Association (NFPA). Their label consists of squares arranged into a "diamond" shape. Each square has a different color to represent the type of hazard. Red represents flammability, and yellow represents reactivity. Blue is used for health hazards, and white is for special information. A number on a square indicates the severity of the hazard. Numbers range from 0 to 4, with 4 representing the greatest hazard. However, any category rated 2 or higher should be considered potentially dangerous.

For example, an NFPA label for toluene would have a 3 in the red square, a 2 in the blue square, and a 0 in the yellow square. Toluene is a chemical used in many paints, paint thinners, lacquers, and adhesives. It is flammable, can be harmful if inhaled or swallowed, and is not reactive when mixed with water.

RED
Flammability
Hazard

BLUE
Health
Hazard

YELLOW
Reactivity
Hazard

WHITE
Special
Hazard

Fire Safety

In order for a fire to burn, three basic things must be present.

- A source of heat
- Oxygen
- Fuel

For example, what if a piece of paper comes in contact with a hot wire and the paper burns? The hot wire provides heat, the paper is the fuel, and oxygen is present in the air.

Most fires can be extinguished by

Reducing the heat The most common way to reduce the heat of a fire is to throw water on it. This has cooling action and produces steam. The steam helps to exclude oxygen. However, water should not be used to extinguish some fires, such as grease or electrical fires.

Preventing oxygen from reaching the fire A fire can be deprived of oxygen by spraying it with an inert gas, such as carbon dioxide. Carbon dioxide is contained in many fire extinguishers.

Keeping oxygen away is also the method used when a person's clothing catches fire. The person is wrapped in a blanket, which smothers the fire.

Removing the source of fuel If a gas or liquid, such as gasoline, is feeding a fire, it can often be turned off in some way. That removes the fuel. In some cases, the fuel may be allowed to burn until it is used up and the fire goes out.

Class of Fire		Type of Flammable Material	Type of Fire Extinguisher to Use	
Class A		Wood, paper, cloth, plastic	Class A	
			Class A:B	
Class B		Grease, oil, chemicals	Class A:B	
			Class A:B:C	
Class C		Electrical cords, switches, wiring	Class A:C	
			Class B:C	
Class D		Combustible switches, wiring, metals, iron	Class D	
Class K		Fires in cooking appliances involving combustible vegetable or animal oils and fats	Class K	

Developing a Fire Emergency Plan

Your lab should have a plan for use in a fire emergency. If your teacher has not explained it to you, ask about it. Be sure you know where all the exits are, where the fire extinguishers are kept, and what your responsibilities are in case a fire occurs. Then follow these steps to record the fire emergency plan.

1. Draw a floor plan of the lab.
2. Locate all the fire exits and label them on the floor plan. If the lab has no direct exit to the outside, draw a map of your section of the school. Show the location of the lab, and draw arrows from the doors in the lab to the nearest fire exits.
3. Locate and label any windows that could also be used for escape. (Check to be sure the opening will be large enough for an adult to pass through.)
4. Locate all the fire extinguishers and label them on the floor plan. Indicate on the plan the class of fire for which each extinguisher can be used.
5. Locate places where flammable or explosive materials should be stored and label them. Be sure they are in a location far away from any source of heat and that their cabinet is fireproof.
6. Determine where you and your classmates should meet after you have left the building. This is important so someone can check to be sure everyone has escaped safely.
7. Ask your teacher for procedures to follow if a fire occurs. If time allows, this may include such things as closing windows, turning off equipment, and grabbing the first-aid kit. Someone should be responsible for reporting the fire. However, human safety is the most important consideration. Fires can spread quickly, and smoke can be just as deadly as flames. Your most important responsibility is to get out alive.
8. Be sure everyone has a chance to study the emergency plan. Then, with your teacher's approval, post it in a prominent place where it can be seen easily.
9. Be sure that you and your classmates are informed as to housekeeping duties that help prevent fires, proper handling of any materials that could catch fire, how to treat burns, and where to report a fire.